CAROLINA BAYS

CAROLINA

PHOTOGRAPHS BY Robert C. Clark

TEXT BY Tom Poland

BAYS

Wild, Mysterious, and Majestic Landforms

THE UNIVERSITY OF
SOUTH CAROLINA PRESS

Publication of this book is made possible in part by the support of the
Harry Hampton Memorial Wildlife Fund: http://www.hamptonwildlifefund.org.

Published by the University of South Carolina Press
Columbia, South Carolina 29208

www.sc.edu/uscpress

Manufactured in the United States of America

29 28 27 26 25 24 23 22 21 20
10 9 8 7 6 5 4 3 2 1

Library of Congress Cataloging-in-Publication Data
can be found at http://catalog.loc.gov/.

ISBN: 1-64336-056-0 (cloth)
ISBN: 1-64336-057-7 (ebook)

frontispiece: *Fog Rising,* Jones Lake, Bladen County, North Carolina

To the memory of Rebecca Sharitz, wetlands ecologist, who guided us through many a bay at the Savannah River Site, and for Linda Lee, wetlands ecologist, who answered our questions and guided us through bays at Savannah River Site as well.

To the memory of my mother, Ruth Walker Poland, who more than once said, "Don't you think you have written enough about these Carolina bays?"—TOM POLAND

Contents

Foreword

Carolina Bays . . . Mystery Solved

We all love a good mystery. We are driven by primal instinct to ask why, how, where, and myriad other questions aimed at solving the mysteries that both plague and enrich our lives. Carolina bays are the embodiment of a good mystery. Since their initial description in 1848, when South Carolina State Geologist Michael Tuomey noted their unique shape and orientation, myriad scientists have been fascinated by these features. Tuomey's work cracked the door open to the mystery of Carolina bays, but the advent of aerial photography in the 1930s blew the door off entirely. Since their early discovery and description, they have both intrigued and bewildered us. In fact many early descriptions labeled them "mysterious Carolina bays," leaving no doubt that our understanding of these phenomena was greatly limited.

Humans encountered and began describing Carolina bays long before their formal discovery. Native Americans made camps along the sandy rims and edges of them. Early explorers and naturalists mentioned them in their writings, giving them their first "unofficial" name: *pocosin.* The word *pocosin* derives from an Algonquin word meaning "swamp on a hill"—and there the mystery begins. The early explorers of our country were accustomed to swamps along rivers, streams, large lakes, and coastal tidelands. Finding a swamp while crossing great stretches of upland was something quite different. No one seems to be sure who originally coined the term "Carolina Bay," but it may have been the early naturalist John Lawson, who in the 1700s noted the abundance of bay trees found in these "swamps on a hill." So even the name, which many associate with an embayment of some sort, is a bit mysterious and may originally have had nothing to do with the embayment or impoundment of water.

It wasn't until the advent of aerial photography in the 1930s that the extent of the real mystery associated with Carolina bays became obvious. Yes, we had read the

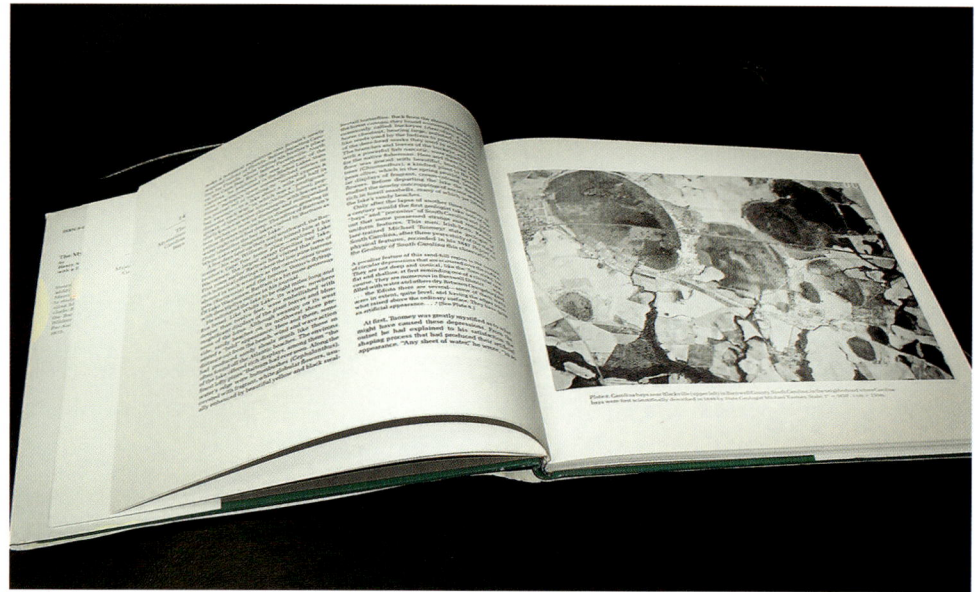

Aerial Photos Kicked Off the Mystery
Aerial photographs of Carolina bays from Henry Savage, *The Mysterious Carolina Bays* (1982), depict the bays' elliptical shape and northwest-to-southeast orientation.

RIGHT **Water on High**
AIKEN COUNTY, SOUTH CAROLINA
Perched water at the Janet Harrison Heritage Preserve.

descriptions of Carolina bays offered by Tuomey and other early researchers, but seeing is believing. Early aerial photos, many from the coast of South Carolina, revealed both great and small elliptical and oval-shaped features spread across the landscape.

And as if to enrich the mystery further, these ellipses and ovals all pointed in the same direction: technically speaking, their long axes were all aligned in a northwest-southeast direction. Some of these features had sandy rims outlining their circumference; some did not. Some appeared to overlap other bays, as if they were stacked one upon another. There it was: visual proof that the mysterious Carolina bays were real.

In 1987 a publication of the University of Georgia's Savannah River Ecology Laboratory included a bibliography of 350 publications related to the study of and research on Carolina bays. In the thirty years since this publication, I believe it is fair to say, numerous additional references to the Carolina bay literature have been added. Early on the vast majority of research on Carolina bays focused on their possible origin and their unique geomorphology—in essence their shape, orientation, and other "physical" features. The theories of origin proposed to date range from far-fetched, such as the one that suggested they were the "wallows" of ancient seagoing creatures, to extraterrestrial, blaming meteor and comet collisions for their presence, to those that focus on earth-bound explanations.

Amphibians' Haven

AIKEN COUNTY, SOUTH CAROLINA

The swollen trunks are a response to the coming and going of water. The absence of long-standing water prevents predatory fish from getting established, a benefit to frogs and other amphibians. Amphibians such as the threatened frosted flatwoods salamander lay eggs only in dry ponds, which guarantees no fish will eat their larva when rains refill the bay.

But this mystery fascinates me less than the ecology of these unique wetlands does. In the past few decades, much of the research on Carolina bays has focused on the plants and animals and unique habitats they support. And in order to understand this mystery, we need to know a bit about the physical function of a Carolina bay—in this case the hydrology or water cycle. Carolina bays are basins that hold water perched above the normal water table due to some sort of confining layer such as clay. Most Carolina bays fill during rainy periods and dry out during dry periods, typically in late summer and fall. This characteristic throws them back into the eco-semantic torrent many of us had to navigate as we went about our studies and research. And there most Carolina bays are functionally similar to an entire suite of wetlands that go by such names as "high ponds," "wet weather lakes," "flatwoods ponds," "bays," "vernal pools," and other somewhat colloquial designations. From an ecological perspective, we have tried to fit them all under the big, and somewhat porous, term "temporary, isolated freshwater wetlands."

The periodic nature of the Carolina bays' water cycle is the key to the mysteries that intrigue me. Numerous species of amphibians, frogs and salamanders, breed either exclusively or preferentially in temporary ponds, including Carolina bays. Because most of the bays fill and dry on some cycle, they seldom have predatory fish present to eat the amphibians' eggs and larvae, and even if fish do get in during particularly wet periods, they disappear as soon as the bay dries out. The same goes for other waterborne predators such as dragonfly larvae. Along with the amphibians, my favorites, these small ecosystems support numerous species of birds, reptiles, other wildlife, and native plants. And a number of these species are quite rare.

So, while I can't solve the mystery of where they came from, I do understand, somewhat, the mystery of Carolina bays as it pertains to their ecological role and their importance in our landscape. We can still discuss and debate their origin in our spare time.

STEPHEN H. BENNETT

Acknowledgments

We thank the following people and organizations for their help in making this book a reality: Helen Clark, the late Ruth Clark, Michael Davias, Jamie Dozier, Bill Funderburk, Amy Geer, Grand Bay Wetland Education Center, Scott Harder, George Howard, Linda Lee, Bob Perry, the late Rebecca Sharitz, Dr. Jim Luken, Chris Moore, Dr. John Nelson, and Johnny Stowe.

Herpetologist Steve Bennett helped identify the plants and animals in several captions over the Internet with messages containing images of the relevant species.

Introduction

Like Going to Africa

Our introduction to Carolina bays came long ago when we worked for the South Carolina Wildlife and Marine Resources Department, now known as DNR, the Department of Natural Resources. I worked in film, and Robert worked in photography.

Robert's first trek into a bay took place in 1983. He had never heard of Carolina bays before joining the Wildlife Department. Steve Bennett, a herpetologist, came into his office and told him there was an incredible place he had to visit.

"I didn't give it much thought," said Robert. "We drove to Dalzell, South Carolina, where we parked our vehicle along a dirt road, Cannery Road. Walking through a sand rim, we soon came to the interior of a pond cypress bay. I saw what looked like hundreds of acres of pond cypresses and grasses so lush they looked unreal. The beauty was overwhelming, and I struggled to get a perspective that would show the beauty I was experiencing. Steve called it 'Dalzell Bay,' but I nicknamed it 'Dazzle Bay.'"

Like Robert, I had never heard of Carolina bays, which were right here in our backyard. My introduction involved a 1981 flight in a Cessna 180 over the bays of Horry County, an appropriate and historic choice, as you'll see. I was to write a script for a natural history documentary, *Mysteries of the Carolina Bays,* and it was critical to get aerial footage. Holding heavy sixteen-millimeter Arriflex cameras, we leaned out the window and filmed the bays below. They were more than mere depressions; they were aligned, elliptical landforms.

Unlike Robert, from an airplane I could not experience the intense beauty of that alien world up close, but I would. For the documentary I made many trips to Woods Bay State Park and Heritage Preserve near Olanta, South Carolina. Here are random handwritten notes from one such excursion.

"Spring, mid to late morning. The wind coos in the pine tops and birdsong surrounds me. The buzz of flying insects gives way to frog songs. Some frogs do, indeed, croak.

Dazzling Dalzell

SUMTER COUNTY, SOUTH CAROLINA

Lush grasses carpet Dalzell Bay's interior, creating a majestic setting few see, a setting of cinematic grandeur. Call it "Dazzle Bay."

RIGHT *Like a Peacock Feather*

BAMBERG COUNTY, SOUTH CAROLINA

LiDAR reveals bays adjacent to Cathedral Bay (blue). These elliptical formations' consistent shapes and northwest-to-southeast orientations amaze, mystify, and fuel theories. These LiDAR images resemble a peacock's plumage. *Image courtesy of Michael Davias*

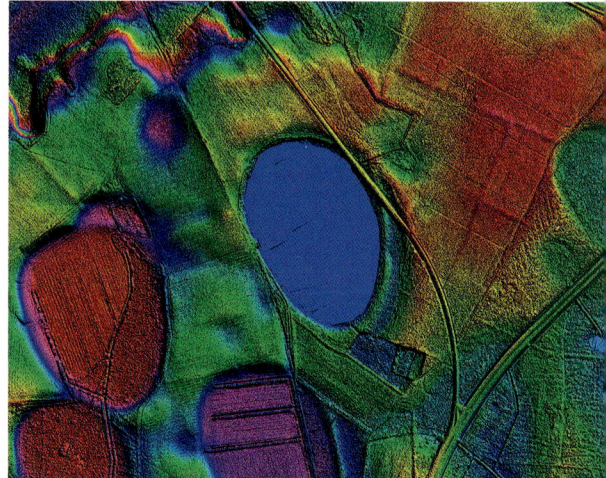

Named for a Town
SUMTER COUNTY, SOUTH CAROLINA

The beauty of Dalzell Bay in Sumter County, a bay Steve Bennett and John Nelson named when they researched their pioneering 1991 *Distribution and Status of Carolina Bays in South Carolina.* They named the bay for the nearby town.

Others bark and some boom. All together it is an amphibian ensemble. My footsteps betray my presence. As I approach, frogs leap into the water. Critters pucker the water from below, spreading ripples across the black, mirror-like water. Blue jays pierce the air with their calls, as raging caterpillars, black with eleven pair of yellow dots, devour leaves. Woodpeckers cry out and hammer a cadence rivaling tribal drums. Sounding his sore-throat squawk, a great blue heron flushes. Startled skinks hurdle into the water where they lift their head high, then skim across the blackwater as if they're wearing skis.

Green on Green

AIKEN COUNTY, SOUTH CAROLINA

The American green tree frog can be heard almost a mile away. They frequent Carolina bays. Usually bright green, their color can vary depending on surroundings, mood, and temperature. The white or pale yellow line running from under the eye to the groin identifies this frog.

North America's Largest Heron

BLADEN COUNTY, NORTH CAROLINA

Widespread and adaptable, the great blue heron stalks bays that hold water year-round. Primarily a fish eater, it has been known to eat most animals within striking range. In flight the great blue brings to mind the ancient pterodactyl.

American Alligator

LOWNDES COUNTY, GEORGIA

Denizen of Grand Bay complex, Valdosta, Georgia.

"It's jungle-like. If a man found himself deeply tangled inside the matted, twisted vines interlocked with shrubs that form long, thick walls of greenery, he just might die before he could claw his way out. The close-packed vegetation merits the word 'impassable.' . . . Just saw a bird that looks like a robin, but its call sounded like that of a bird in the Congo basin, like a Hollywood sound effect. From afar, a bellowing gator sounds like a tiger growling in some tropical jungle. I've transported myself into Joseph Conrad's *Heart of Darkness*."

And that was just one morning in one bay.

As we explored bays, we kept encountering sights that stirred up images of the Dark Continent. Here's a note scribbled onto a legal pad in Florida Bay, known also as Wambaw Bay: "In the savanna, wind-sculpted pond cypress tops evoke Africa's acacia. I half expect to see a cheetah bounding through the grasses." The U.S. Forest Service cored these trees and determined they are more than four hundred years old. They're not that tall, but the trees have history and character.

CLOCKWISE FROM TOP LEFT

Wind-Sculpted, Fall Version
In a bay's savanna, trees often have twisted, flattened canopies nowhere near as symmetrical as their forest counterparts. Trees' closer spacing in forests provides some buffer against the wind. Trees in savannas, however, yield to the wind's power.

Twisted Trunk
Strong winds misshaped this pond cypress at Florida Bay.

Foresters Convene
U.S. Forest Service personnel examine the savanna in autumn at Florida Bay, known also as Wambaw Bay. The bay, off Highway 45 in the Lowcountry, is part of the Francis Marion National Forest.

And here's an observation of Red Bluff Bay: "Colonies of pitcher plants loom over the grasses like alien cities. Surely, if life exists elsewhere in the universe, this is what it looks like." Their unusual blooms appear otherworldly as well.

Humanity has a long history with bays. People have interacted with Flamingo Bay for thirteen thousand years. They still do. At Savannah River Site's Rainbow Bay, the University of Georgia's Savannah River Ecology Laboratory's "frog watch" set a Guinness world record in 1999 as the world's longest-lived daily amphibian study.[1] Carolina bays astonish us in many other ways.

We explored approximately thirty Carolina bays scattered across South Carolina, North Carolina, and Georgia. Some have regal names—a few, to wit: Cathedral Bay, Emerald Bay, Rainbow Bay, Thunder Bay, and Florida Bay. Some, such as Cartwheel Bay, have names that make you scratch your head, and some take their names from people: Craig's Pond, Savage Bay, and Bennett's Bay, honoring our friend Steve Bennett. Several take their name from Indian tribes—Lake Waccamaw, for instance. Some, such as Dalzell Bay, take their name from nearby communities. In Cook County, Georgia, you'll find No Man's Friend Pond, a Carolina bay whose name hints at the difficulties and challenges bays present.

Extraordinary Trio
BERKELEY COUNTY, SOUTH CAROLINA
Full sun, ample water, and poor soil supplemented by insects provide carnivorous pitcher plants habitat. Blossoms as unusual as the pitcher plant itself result come spring.

Rainbow Bay's Record
Linda Lee holds Rainbow Bay's Guinness certificate.

RIGHT **Haunted Waters**
Autumn. Carolina bays offer surreal beauty throughout the seasons.

All these years later, we know a bit more about the mysterious landforms Indians called pocosins—"swamps on a hill." Outside select geologists, geomorphologists, botanists, herpetologists, biologists, naturalists, writers, and photographers, few know these bays intimately. Right here it is important to make a statement. This book is intended for the public, but it makes reference to science, physics, and theories of origin. We endorse no one theory and understand

that no one theory enjoys scientific consensus. Whatever created Carolina bays remains an enigma.

Mystery aside, many people know nothing about Carolina bays. Many have never even heard of them. So don't feel alone if you're unfamiliar with Carolina bays. You may find it difficult to picture a bay. Don't worry. Our book will take you into them. As you'll see from its photographs, they're hauntingly beautiful. LiDAR photos precisely illustrate that they are elliptical depressions that stretch across the Atlantic coastal plain from Florida to Georgia, the Carolinas, and on across Maryland and Delaware into New Jersey.

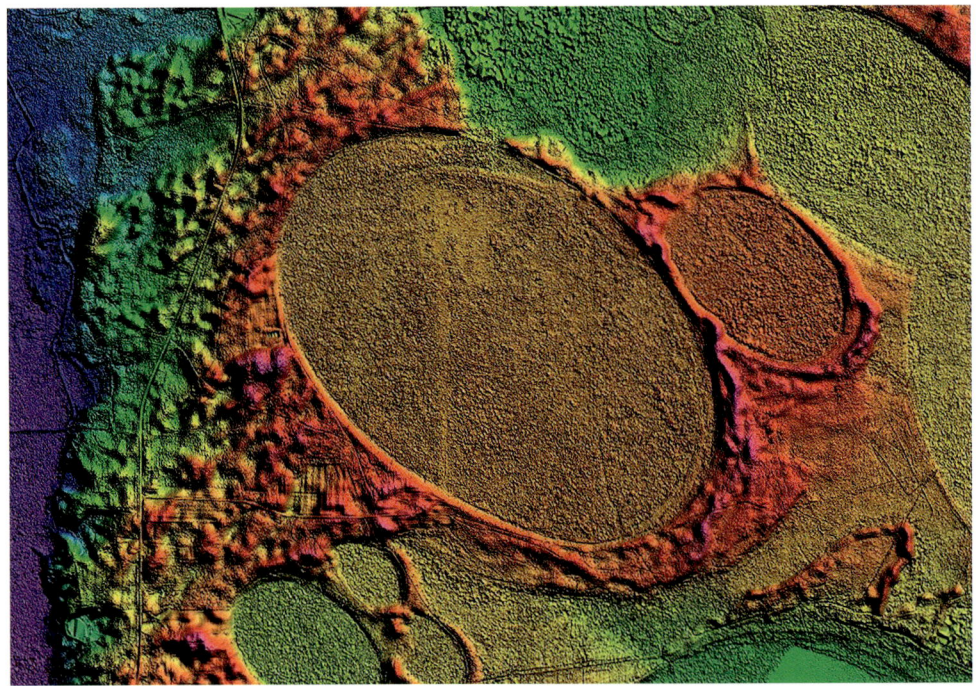

Jerome Bog
BLADEN COUNTY, NORTH CAROLINA
Big Bay in North Carolina and smaller bays clearly reveal the trademark northwest-to-southeast orientation. *Photo courtesy of Michael Davias.*

Marbled Weaver
CHARLESTON COUNTY, SOUTH CAROLINA
A marbled orb weaver rests dead center in its web. A member of the genus *Araneus*, it's a common spider. Despite its fierce appearance, it's nonvenomous. Bays' abundant insects offer prime food supplies to spiders.

Thousands, perhaps a million, dimple the coastal plain. Elliptical depressions of varying sizes, they align northwest to southeast. To fly above them is an eye-opening experience, but even more astonishing is exploring them on foot. We have done that and herewith share our efforts to document these otherworldly places where wilderness and nature provide a balm to civilization's ills.

A balm, for sure, but that hasn't stopped people from draining thousands of Carolina Bays and converting them into farmland. Other bays have been lost to highway

Floral Masterpiece
CHARLESTON COUNTY, SOUTH CAROLINA
The showy yellow-fringed orchid grows in moist, acidic soils in the pine savannas of Carolina bays. Butterflies find *Platanthera ciliaris* highly appealing. American Indians found medicinal purposes for the plant.

construction, housing developments, and even golf courses. The undisturbed bays that remain deserve our understanding and protection, for beauty and purpose fill them, and all that beauty and purpose exists for a simple reason. People have left them alone.

Robert and I used to be among those who had never heard of the bays, but careers at a wildlife department introduced us to them. Many years later we explored them on our own. It was hot. It was steamy. We took spills in blackwater and suffered wasp attacks and other hardships. We logged more than ten thousand miles across North Carolina, South Carolina, and Georgia. Now we bring the bays to you. We believe you will find them amazing and worthy of preservation and, in the cases of altered bays, restoration.

We are not naturalists; we aren't scientists, nor should you consider this book an academic exercise. We appreciate what bays do, and we hope this book brings more attention to Carolina bays' beauty, significance, and benefits and the need to reclaim those we can while preserving the few left undisturbed. These pockets of true wilderness remind us of just how much we've lost to housing developments, strip malls, highways, pipelines, farms, factories, and power lines.

Carolina bays remind us that like prothonotary warblers, spotted salamanders, green hooded pitcher plants, yellow-fringed orchids, and marbled weavers, we, too, sprang from a sublime natural world. Never forget where you come from, they say. It remains sound advice in the twenty-first century *Anno Domini.*

More sound advice comes from James Dickey's "East of Eden," which first appeared in 1989 as the foreword for *South Carolina, the Natural Heritage,* our book about the need to protect undisturbed natural areas. Its relevance to Carolina bays cannot be overstated. As well, "East of Eden" reunites the team of Steve Bennett, Robert Clark, James Dickey, and Tom Poland a final time.

"East of Eden"

We are all in danger of becoming outsiders, thrust by our own technology "East of Eden"; displaced persons who can no longer reach in any personal way the colors on the trees, the slow turning of the earth through the seasons. It has been over a hundred years since a poet—George Meredith—could say, "my heart shot into the breast of the bird," and mean it.

And yet there is a return, so long as the world lasts, so long as the plants and creatures hold their life.

For it is life—*life*—they hold—nothing else—these leaves and flowers and creatures, among whom we perpetually find ourselves, exiled in mind as we are. Any time we turn our eyes to the natural world we are moving along that return, and into a kind of personal vision: the world itself as revelation, and one given only to the living. In our time we must ask ourselves how much of this we are damaging, and ask also whether anything we have created is so beautiful or original as the least aspect of the vision bequeathed to us. Philip Wylie, for example, would have us believe that "the sum total of human works, the artifacts of savages, barbarians, medieval men and modern, all cities and towns, every hut, hovel, skyscraper and temple, all steel bridges, everything man has made to use since the first stone tool or wooden club—does not equal, in all parts put together, the achievements of the life forms of plant and insect in a square foot of grass." Yes; it is true. What laboratory could make that square foot? Even if the scientists could contrive the grass, it would still be an imitation, strictly second-hand, inferior. Would it not?

During any encounter with a part of the God-created cosmos it is good to remember the word "emergence." If the mind blanks properly and the instincts take over, one will sense something like this happening, as the self drops off and the feeling part of the

Red-Bellied Woodpecker
In the bays red-bellied woodpeckers pick at bark and fly from tree to tree with an undulating flight, rising and falling, dipping then descending.

RIGHT **Grand Bay Stalker**
LOWNDES COUNTY, GEORGIA
Bays bring open spaces and beauty to Georgia, South Carolina, and North Carolina. Note the heron in the lower right-hand corner near Grand Bay in Georgia.

personality—spirit, what used to be called the soul—enters into union, "like a sudden extension of consciousness," with the thing beheld: the entity small or large, fleeting or permanent, the snowflake or the galaxy, the leaf or the nebula.

South Carolina has an abundant share of these natural visions; all we need to do is to enter into the states of being they make possible: as the naturalist John Hay says, each state constitutes "a first opening into original space." From the mountains of Oconee to the beaches of Pawleys Island, these kinds of original space exist any time we open ourselves to them, and merely *look* without thought of using: of subdividing, of putting up factories and shopping malls, of mining, of extracting, of gathering, of any kind of exploitation. We need to substitute instinct for reason, and merely to *behold* what is; and

view with primal innocence those parts of creation that have nothing to do with us except to serve as material for contemplation and wonder, born into it as we were, living by means of its processes, part of a mystery, part of a whole scheme, a cosmos, the reason for which will never be known. I would like to believe that this book may be for certain human beings the true opening of some of these windows, in our State, into these possible states of existence, as though the reader were being prepared to walk, as he should, bareheaded, open-faced, and step-by-step, into a miracle, saying, "*It* is here. Still. And *I* am here," but returning always to the words, "*It* is here. Still. While I am alive. It is still here."

James Dickey

OVERLEAF **Up from Black Waters** BAMBERG COUNTY, SOUTH CAROLINA

The Big Oval Picture

Before we delve into a Carolina bay's features and functions, let's spend time in one. Abandon civilization for a place that's wild, natural, and unique.

Late afternoon. Buttressed cypresses uphold an impenetrable canopy, and yet sunlight slants across blackwater, throwing cypress knees' shadows afar. Spanish moss glows as motes of organic matter sparkle in shafts of light. Reflections shimmer across coal-black waters, as wind surging through treetops sounds like falling surf, an

Seasonal Change

LANIER COUNTY, GEORGIA

Come fall, the vivid hues of spring and summer give way to earth tones. From the tops of cypress to the grasses and pitcher plants, subtle tones hint that winter is coming to the bay. Spanish moss? It seems unaware of the seasonal change.

Yellow Jessamine
SUMTER COUNTY, SOUTH CAROLINA
Along a bay's sand rim, you'll find the South
Carolina state flower, the yellow jessamine,
known also as Carolina jessamine. Toxic to hu-
mans and wildlife, even deer refuse to eat it.
Its yellow clusters of blooms remain, however,
a feast for the eyes.

Anhinga Takeoff
BLADEN COUNTY, NORTH CAROLINA
At Jones Lake—a "lake" that's the cumula-
tive result of rainfall—an anhinga takes flight
through filaments of fog that rise from cool
waters.

in-and-out subsiding of silence. A short walk away, winds ghost across the savanna. A
walk along the sand rim feels beach-like, but alien plants claim these sands, as do familiar
ones like the yellow jasmine.

Find a spot and sit as still as a stone. Be quiet. Blend into your surroundings, and
the bay comes alive. Birdsong fills the air. An anhinga, the snakebird, dries its wings.
A gator glides past water lilies; a cottonmouth slides over leaves. Mallards land with a
glassy crash of water. Turtles rest on logs. You see an abundance of wildlife: ospreys, carnivorous plants, waterfowl, and otters. And there's the profusion of colors wild animals,

birds, insects, amphibians, grasses, orchids, and other living things display. Few things equal the beauty of a cathedral-like bay. In a Carolina bay, it's easy to imagine you are in some faraway world.

Solitude. That word applies to Carolina bays. No, two words, both with equal weight. Solitude and wildness. Wildness, for a Carolina bay is not a park. It's wilderness. Beautiful, significant wilderness, but debatable as origins go—that's Carolina bays.

So how do you introduce what some authorities believe may well be the world's most mystifying landform? Let's begin with the basics. Bays are shallow, elliptical depressions of uncertain origin in the southeastern coastal plain. They formed mainly in level landscapes made up of sand, sandstone, limestone, and other sedimentary materials.

Consistent positioning is a bay trademark. Their long axis runs northwest to southeast. This orientation gives them a parallel appearance. The larger bays possess rims of white sand on their southeastern end, sometimes identified by the unknowing as "ancient beaches." Some sand rims stand fifteen feet high; widths can extend several hundred feet. Often the sand is as white as snow. Thick layers of soil and peat deposits characterize many bays.

Bright Thrush of the Bays
CHARLESTON COUNTY, SOUTH CAROLINA
Frequently burned savannas were among eastern bluebirds' original habitats. They find bays' sand rims and high shrubs to their liking.

A Desertlike Place
DILLON COUNTY, SOUTH CAROLINA
Sand rims on bays' southeastern edges form a xeric, desertlike environment. The rims support pines and scrub oaks, but it's a hardscrabble existence. The white sands resemble a dusting of snow, and in autumn the reds of turkey oak leaves contrast sharply with the beach-like sands.

Plenty of bays are bogs or periodic wetlands important to migrating waterfowl. Herons, egrets, and wood storks frequent them, as do dragonflies, green anole lizards, and green tree frogs. Isolated wetlands, they sometimes overlap or occur as a smaller bay inside a larger one. They vary from a few square feet up to thousands of acres. North Carolina's Lake Waccamaw is a Carolina bay 5.2 by 3.5 miles. At 8,938 acres it's one of the larger bays. Local Indian legend says that a bright ball falling from the heavens created Lake Waccamaw.[1] As bays go, it is unique in that it's spring fed and forms the headwaters of the Waccamaw River. Georgia's Old Field Bay sprawls 6,000 acres in the Grand Bay Wildlife Management Area. Twenty-five bays in South Carolina feature long axes longer than 10,000 feet.

As to how many exist, estimates run from five hundred thousand to as many as a million thanks to new surveying technologies such as Light Detection and Ranging (LiDAR, among other acronyms), which measures distance by illuminating a target with a laser light. Approximately seventeen thousand bays of various sizes dimple South Carolina's coastal plain, according to research by Silvia Piovan and Michael Hodgson.[2] Jerry Reynolds, Carolina Bay authority at the North Carolina Museum of Natural Sciences,

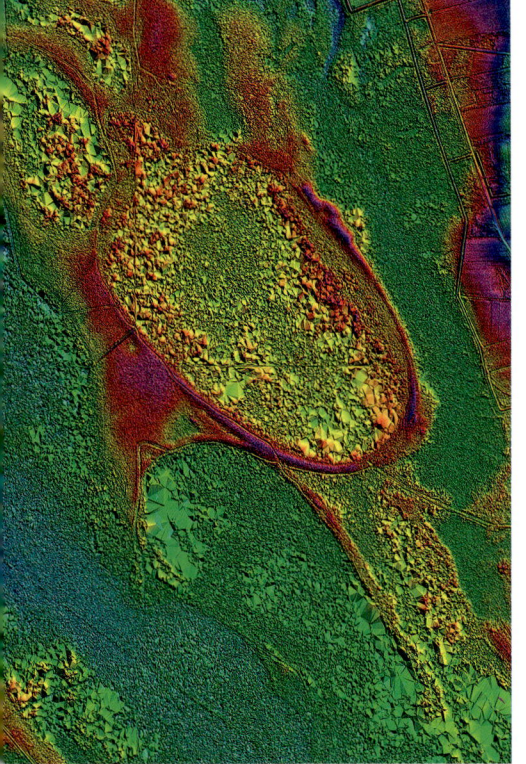

Lit Up by Laser
HORRY COUNTY, SOUTH CAROLINA
Cartwheel Bay. Light Detection and Ranging (LiDAR) surveying technology illuminates a target with sixteen thousand laser light pulses a second. The pulses penetrate canopies and measure elevation, letting researchers build digital elevation models with precise range, incident angle, and surface composition. An excellent tool for mapping and studying bays with precision, its creations are quite colorful. *Image courtesy of Michael Davias*

RIGHT *Darwin's Wondrous Plant*
HORRY COUNTY, SOUTH CAROLINA
A carnivorous plant, the Venus flytrap evolved to survive in nutrient-poor soil. Small and lovely, it's one of the plant world's more ingenious predators.

has seen estimates ranging from four hundred thousand to 2.5 million bays in the Southeast.[3]

We don't know what created them. We don't even know how many exist, but what ancient wonders they are, though determining their age is problematic. A variety of dating techniques place them at thirty thousand to one hundred thousand years old or older. No "new" bays appear to be in development today, although the possibility exists that certain high ponds could be bays in development.

Maybe.

Much remains to be known, as origins go, for Carolina bays wear a badge of mystery. Two major questions need answers. What formed the bays, and when did they form? The origin debate is ongoing. As for when, ages range all over the chart. Depending on which researcher you read, ages range from 7,000 to 200,000 years. Certain ages occur more often than others: 12,900 and 45,000 years. As closed ecosystems, bays have been around a long time. There's very little to agree on except this: Carolina bays are fabulous but fragile landforms that contain a rich diversity of habitats.

South Carolina's research role in bays is an important one. State geologist Michael Tuomey first recognized Carolina bays in 1848 during a geological survey in South Carolina. Wrote Tuomey, "A peculiar feature in the topography of this sand-hill is the number of circular depressions that are scattered over the surface. They are not deep and conical, like 'lime-sinks,' but flat and shallow, at first sight reminding one of a circular race-course."[4] Whether he intended to or not, Tuomey may have well started the origins controversy: "The circular form of these depressions is easily accounted for. Any sheet of water, in so incoherent a soil as that covering the surface of this region, however irregular its outline, would soon become circular, because the projecting irregularities would be washed away."[5] Geologist Douglas Johnson wrote in 1942, "The Carolina bays are without doubt one of the most remarkable geomorphic features on the surface of the earth. They share with submarine canyons the distinction of being among the most difficult of earth forms to explain."[6]

Origin theories excite and incite. For well over 170 years, researchers have dueled over theories attempting to explain the bays' presence. Nineteen origin theories have been set forth. Scientists fall roughly into two camps: extraterrestrial theorists, who turn to meteorites and comets as explanations, and those who believe natural processes such as wind and water formed the bays. Their origin remains the topic to this day.

Let's return to a chilly January morning in the sprawling confines of Savannah River Site (SRS). Robert Clark, wetlands ecologists Linda Lee and Rebecca Sharitz, and I are driving around ice-felled pines on our way to explore several bays, a scouting trip for photography to come in the spring. At SRS sixty percent of the rare and threatened plant species live in Carolina bays.

Launching into a discussion of Carolina bays, it doesn't take long until the question comes up. "So, Tom, what do you think is behind the bays' origins?" asks Sharitz.

"Well, it's not the meteorite theory," I reply.

"Good," says Sharitz.

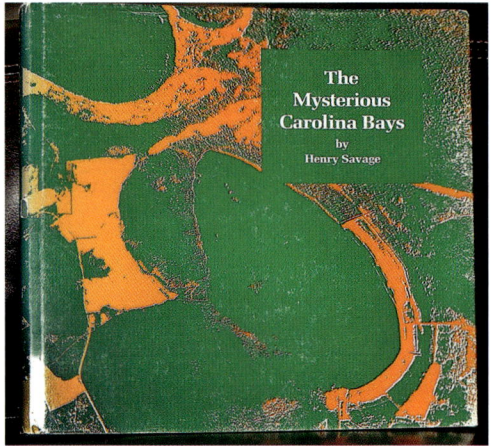

Savage Bracts

KERSHAW COUNTY, SOUTH CAROLINA

Dogwoods' snowy bracts (modified leaves, not blossoms) brighten Savage Bay, a Carolina bay named for Henry Savage, a Camden, South Carolina, naturalist and attorney who published *The Mysterious Carolina Bays* in 1982.

RIGHT **The Mystery Endures**

Camden attorney Henry Savage advocated the meteorite theory as the origin of Carolina bays, a theory that has been proven to be untrue.

Right away we establish common ground, but others debate theories that range from outlandish to probable. One theory claims dinosaur footprints created the bays. An icy comet blasted them out. Throw in ancient volcano activity. Giant turtles hollowed them out. No, ancient buffalo wallows formed them. Not true, fish fanning their fins in unison over submarine artesian springs created them. Sinkholes like we see in Florida's limestone solution areas created them. And on it goes.

South Carolina can lay claim to two theories of origin. In 1977 Ray T. Kaczorowski developed a thesis, "The Carolina Bays and Their Relationship to Modern Oriented

Lakes," while working on his Ph.D. in geology at the University of South Carolina.[7] His thesis led to a book of the same title, published by University of South Carolina Press.

Henry Savage, a Camden naturalist and attorney, had a book published in 1982, *The Mysterious Carolina Bays* (University of South Carolina Press). Savage's theory that meteorites created the bays got a lot of attention. Though his theory never took hold, his work as a naturalist led to the naming of a bay in his honor, Savage Bay, a heritage preserve featuring rare plants such as Georgia bear grass and more common plants such as dogwoods.

The meteorite theory proved most exciting and remains part of the bays' allure. Just imagine a bombardment of meteorites gouging out elliptical bays across the Southeast coastal plain. The fact is, scientific consensus on how and what formed bays remains elusive.

An Unfortunate Name

Consider the name "Carolina bays." The most and perhaps the best bays are concentrated in the Carolinas, but they occur from New Jersey to Florida and Alabama and are not connected to the sea, bays, or even creeks and rivers. "Carolina bays" is a misleading name. Ask someone who knows little if anything about them, and they'll guess that the bays have something to do with coastal waters. Well, they're not ocean inlets. Few bays contain water. *Bay* refers to the preponderance of bay trees, especially sweet bay, red bay, and magnolias and laurels that dominate them. (The bay leaves that flavor spaghetti sauces come from laurel bays.)

It's quite interesting that the first use of the word *bay* was in *Science* in 1895, when L. C. Glen described Carolina bays near Darlington. "Having noted on a surveyor's map of my school district of Darlington, S.C., several lake-like expanses usually represented as being at the head of some very small stream, I began inquiries concerning them and followed this up by visiting several of the largest. To the lake-like expanses the term 'bay' is usually applied and by it is meant a perfectly flat, clay area with a surface some two to four feet below the general level of the country. . . . Growing in this water, where the 'bay' is uncleared, are cypress, juniper or black-gum trees with a moderately thick swamp undergrowth." Glen went on to write, "A sand ridge borders each 'bay' on the east and southeast."[8]

In 1933 Frank A. Melton and William Schriever of the University of Oklahoma coined *Carolina bay* in the first publication to suggest an extraterrestrial origin.[9] Bays have other names. They're known as Delmarva bays in Delaware and Maryland as well as Maryland bays. They're also called Maryland basins. Other landform depressions,

Autumn Sand Rim
Sand, scrub oaks, and pines—autumn's beauty
touches the southeastern edge of a bay.

LEFT **Carolina Bay Still Life**
BERKELEY COUNTY, SOUTH CAROLINA
Bay trees gave Carolina bays their name in
part, and pond cypresses form part of their
majestic identity. This pond cypress stands in
a savanna, but in many bays pond cypress rise
from buttressed trunks to tower over black
water, casting reflections across the water.

not widely accepted as Carolina bays, exist within the northern Gulf of Mexico coastal plain in southeast Mississippi and Alabama, where they are called Grady ponds or citronella ponds. Nebraska has landforms called Nebraska rainwater basins that appear to be Carolina bays. Bob Kobres, who writes about comets, referred to them as "Lucifer's footprints," as if, indeed, they are impact craters.[10]

What might be a better name? Hard to say. Most bays have names, given our propensity to name everything. Many names are as colorful as the bays: Devils Woodyard Bay, Beeswax Bay, Horseshoe Lake, Misery Bay, Tiger Bay, Bare Bone Bay, Boggy Gully Bay, Blue Moon Bay, and Wolf Bay. Cotton Patch Bay, near Myrtle Beach, is one of the bays that opened the eyes of the world to bays' uniqueness when photographed from the air in the 1930s. Morse Code Bay, in the Savannah River Site, provides an example of the wide-ranging inspiration for bay names. Regardless of their varying names, they share a distinction. They belong to a class of landforms that defy explanation.

Water Collectors

Most bays do not hold water, but those that store seasonal rains do so for a simple fact. Carolina bays are shallow, elliptical depressions with clay or some other impervious bottom. Some absorb groundwater. Their depth ranges from a few feet to as much as forty feet. It's an oversimplification, but think of them as oval saucers lying above the water table sunk beneath surrounding lands. Try this experiment: take a teaspoon and press it ever so lightly into the earth, making all impressions parallel. For greater precision, use a compass and orient your indentations northwest to the southeast. Imagine

Water for a Carolina Bay
BLADEN COUNTY, NORTH CAROLINA
Most Carolina bays fill during rainy periods and dry out during late summer and fall. Think of them as basins set into the coastal plain—basins open to rains, for few are spring fed. Thanks to rain, Carolina bays represent the coastal plain's dominant freshwater wetland feature.

How Many Teeth?

AIKEN COUNTY, SOUTH CAROLINA

Alligator mississippiensis, juvenile. Gators have between seventy-four and eighty teeth. As they wear away, new ones take their place over a lifetime of predation that can wear out three thousand teeth. This species has lived on this watery planet for more than 150 million years, and it's avoided something the dinosaurs could not—extinction. A good place for young gators? Carolina bays.

It's Not a Copperhead

BERKELEY COUNTY, SOUTH CAROLINA

The corn snake, a constrictor that's a species of rat snake, is slender with a belly checkered white and black, like a piano keyboard or Indian corn. Some mistake it for a venomous copperhead. A good climber, it's most common in the coastal plain and favors sandy pinewoods.

The Northern Cardinal

UNNAMED BAY, RICHLAND COUNTY, SOUTH CAROLINA
Birdsong fills bays, and the cherry-red bird adds its chirping, variable-speed trill to the mix. You'll hear its "cheer, cheer, cheer, birdie, birdie, birdie" call. A vocalist, it has sixteen different calls.

the wider northwest end being filled with pond cypresses standing in blackwater. In the middle of the depression, imagine a savanna. At the smaller, southeastern end, visualize a rim of white sand. Put a bit of water in several depressions. Stand over the dimpled clay with these images in mind, and you get a rough notion of how they look from far above.

When winter and spring rains come, the shallow depressions fill with water, but summer and fall dry them out. One season they'll have water; another season they won't. Plants and animals come and go with seasonal rains. Of those that hold water year-round, think of them as isolated basins of water. Even though many Carolina bays hold water only part of the year, they provide important habitat. Throughout the year the bays serve as a haven or refuge for many wildlife species and rare plants. Venus flytraps, pitcher plants, sundews, salamanders, snakes, alligators, wading birds, and deer find support and sanctuary in Carolina bays. Bays are rich with feathers—herons, egrets, and migratory waterfowl flock to them. Raccoons, skunks, otters, and opossums get food and water from Carolina bays. Zooplankton live in them. Salamanders and frogs, especially, are abundant in the bays, which provide breeding sites.

Though it's inaccurate, the name "Carolina bays" means nothing to bays' true residents, their flora and fauna, microbes, and other life-forms. And in a way, the name adds to bays' mysterious nature. Ask a person if they've seen a Carolina bay, then watch bewilderment betray itself on the face of the unknowing.

Pocosins

You can't talk about Carolina bays without mentioning pocosins. The terms *bay* and *pocosin* often appear together. *Pocosin* is one of those strange words that rolls off the tongue of ecologists, biologists, and botanists as easily as a skink dives into water. Ecologists put forth a general definition most experts agree on. You'll find pocosins occurring in broad, shallow basins and on broad, flat uplands with sandy humus, muck, or peat bottoms. While *pocosin* refers to a type of wetland with deep, sandy, acidic, peat soils, it also references an evergreen shrub bog community growing in often wet, nutrient-poor soils, as in bays. The soil may be poor, but pocosins prove to be near-impenetrable thickets. The tangled vegetation results in the thickest undergrowth imaginable.

Aside from pocosins, trees are plentiful in bays. Expect to see pond cypress, black gum, sweet gum, magnolia, bald cypress, and maple and shrubs such as sumac, buttonbush, and gallberry. Blueberries grow in bays. And of course so do the bay trees and laurels that gave them their confusing identity. Water lilies, pitcher plants, sedges, and

A Banquet's Remnants

CHARLESTON COUNTY, SOUTH CAROLINA
Fox squirrels, because of their large size and strength, feast on longleaf cones. As longleaf pines have declined, so have fox squirrels.

Unyielding Green Wall

FLORENCE COUNTY, SOUTH CAROLINA

In Woods Bay State Park and Heritage Preserve near Olanta, South Carolina, beyond the trees waits an impenetrable wall of unyielding green, a pocosin, Algonquin for "swamp on a hill." Pocosins' tangled vegetation, the thickest undergrowth imaginable, is rightly described as impenetrable.

grasses carpet the bays. Near the South Carolina–North Carolina line, a small number of Carolina bays host the legendary Venus flytrap.

Carolina bays are fragrant places spiced with the scent of flowers, orchids, and the oil of cypresses. Sight, smell, and hearing get good workouts in the bays. So have you seen a Carolina bay? Without doubt you have. At ground level most people don't recognize them. When you drive by one filled with water, you think, "Ah, a swamp." Not always true. If only you could rise high above and see its classic oval shape, you'd experience a thrill akin to bays' discoverers. The Ocean Forest Company of Myrtle Beach contracted the Fairchild Aerial Survey Company in 1930 to survey Horry County. Droning along, drawing eyes upward, Fairchild's FC-2 cabin monoplane crisscrossed the coastal plain. What its photographer must have felt when he focused on the mysterious ovals below!

Fairchild's collage of elliptical landforms startled everyone. Depressions dimpled the earth as if an immense chef had pressed spoons into the land anticipating the broth of living things sure to fill their basins. And fill them it did.

Carolina Bays: A Summation

Carolina bays are unique wetlands found in the U.S. coastal plain from southern New Jersey to northern Florida. Optically stimulated luminescence (OSL) dating reveals several periods when bays came to be during the Pleistocene, about one hundred thousand years ago.

All Carolina bays are naturally occurring, shallow depressions that share some of many characteristics: an age of at least thirty thousand years; elliptical or oval shape with northwest-to-southeast orientation of the long axis; low, sandy rims with the larger rims being on the southeast margin; soils of either clays and silts or organic peat that differ from surrounding land; flat, sandy bottoms beneath interior fill; and standing water occurring on a seasonal basis to year-round. The vast majority of bays are independent of inflowing or outflowing waters. Carolina bays are perched, that is, they sit above the water table.

Sizes vary from a tenth of an acre to North Carolina's Lake Waccamaw (8,938 acres). Due to variations in rain and water collection, bays can be inundated for a period, then dry for a period.

Magenta Summer Bloomer

CHARLESTON COUNTY, SOUTH CAROLINA

The grass pink orchid found in bays has no nectar, so it tricks pollinators. Brushy hairs on the uppermost petals resemble the pollen-bearing anthers of other species. This duplicity lures insects, which pick up pollen sacs, allowing pollination to occur.

RIGHT **Moss-Clad Cypress**

Spring greens up a bay.

Containing pond cypress waters, savannas, and dry sand rims, Carolina bays support a diversity of species and are often referred to as "islands" high in species richness. Biological diversity is a hallmark. Among bays' residents are flatwoods salamanders, Carolina gopher frogs, egrets, skunks, herons, turtles, snakes, and alligators. Vegetation is rich in bays. Sweet gums, maples, buttonbushes, gallberry, sumac, and grasses and sedges live in them. The wettest bays have open-water and floating-leaved aquatic plants like water lilies or tall grasses and rushes. Many bays feature an open canopy of pond cypresses. Many provide much-needed breeding sites for amphibians.

Bays are significant for their diverse habitat and rich biodiversity, but many have been disturbed, with the most common causes being logging, ditching, row crops, highway construction, and residential and commercial construction. Urbanization sacrificed Blythe Bay in Wilmington, North Carolina, to a hospital, shopping mall, and golf course. Many sand rims have "borrow pits" where people have taken sand for their own use. Thanks to an unfortunate U.S. Supreme Court ruling, bays are not protected by the Clean Water Act, as they're considered isolated wetlands with no inflowing our outflowing streams. It's hoped this situation will be corrected.

Amazing Landforms

It took a world war and an aerial viewpoint to put Carolina bays in perspective. In World War I airplanes first played a major role in armed conflict. At the war's outset, ground forces dominated battles, but Allied airplanes such as the Curtiss Jenny and de Havilland DH.4 engaged German Fokker Dr.1s and Albatros D.IIIs in dogfights and more. As the Allies acquired superior air strength, they disrupted German ground forces, and the war soon ended. Just like that, aerial observers' sketching and drawing were rendered obsolescent.

Aerial photography proved to be a strategic way to plan battles, a way to see better what the enemy was up to. Aerial photographs helped create superior battle maps, and technology rushed to make aerial cameras specifically for use in flight. Taking pictures aloft, however, wasn't a snap. Stability and shutter speed created difficulties. Then, near the conflict's end, Sherman M. Fairchild developed a camera with the shutter inside the lens. This innovation reduced the significant image distortion created by slow shutter speeds that couldn't keep up with an airplane's movement. Fairchild's design was so good that it became the standard for aerial camera systems for half a century.

After the war Fairchild took overlapping photographs and made an aerial map of Manhattan Island. This aerial map became a commercial success, and deals with other cities followed. Aerial photography proved it could be a successful commercial venture. Aerial photos, for instance, were faster and much less expensive than ground surveys.

Fairchild was onto something, and he wasn't hurting for a pedigree. Born in 1896 in Oneonta, New York, he was George Winthrop Fairchild and Josephine Mills Sherman's only child. His father, a Republican congressman, cofounded IBM and served as its first chairman. In February 1920 Fairchild founded Fairchild Aerial Camera Corporation. Wanting to expand into mapmaking and aerial surveying, he formed Fairchild Aerial

Surveys a year later and bought a surplus World War I Fokker D.VII biplane to take his cameras aloft.

Sherman Mills Fairchild's efforts culminated in a company that would introduce the bays' unique shape and alignment to the world. In 1930 the Ocean Forest Company of Myrtle Beach contracted Fairchild Aerial Surveys to take aerial photos of the area.[1] Fairchild flew over some five hundred square miles of coastal plain in and around Horry County. His FC-2 cabin monoplane canvassed the coastal plain taking aerial photographs. Photograph after photograph revealed unexpected identical landforms. On and on they went, one elliptical depression after another.

Like Craters on Mars

Later, Franklin Roosevelt's Department of Agriculture studied East Coast aerial photography, and a staggering finding surfaced: the egg-shaped landforms ran northwest to southeast. The similarity to craters on Mars and the moon astonished people. Did a celestial bombardment gouge out what would become Earth's most mystifying landforms? One writer described the bays as looking "as if some outraged giant had blasted it with a colossal shotgun."[2]

The discovery triggered immense interest among the general public and researchers. In 1933, just three years after Fairchild's aerial photography, the University of Chicago Press published a fourteen-page paper in the *Journal of Geology* by researchers Frank A. Melton and William Schriever of the University of Oklahoma. What follows is the abstract of "The Carolina 'Bays': Are They Meteorite Scars?"[3]

Aerial photographs of a district on the coastal plain of South Carolina reveal hitherto unknown relationships among surface depressions of a peculiar type, the origin of which has long been a subject of speculation. These relationships include (1) a smoothly elliptical shape, (2) parallel alignment in a southeastern direction, (3) a peculiar rim of soil which, with unimportant exceptions, is invariably larger at the southeastern end than elsewhere, and (4) mutual interference of outline. Consideration of all of these facts leads to the conclusion that the origin is not directly attributable to ordinary geologic processes. On the contrary, a hypothesis involving impact by a cluster of meteorites is presented as the most reasonable explanation. The supposed swarm must have been large enough to possess a cross-sectional area at right angles to the direction of movement of the order of magnitude of 50,000 square miles.

Carolina Bays Is Coined

Melton and Schriever were the first to use the term *Carolina bays,* taking a cue from L. C. Glen, who used the word *bays* in referring to the elliptical depressions in 1895. In stilted language Melton and Schriever anticipated the bewilderment the name would create. "Any undrained depression which contains water throughout most of the year is locally designated by the term 'bay.' In view, however, of the unusual nature of the hypothesis herein presented, the authors prefer to use this word temporarily for the type of basin under discussion in spite of the ambiguity involved. At some future time, if it seems desirable, the more logical appellation 'scar' may be applied, or another name may be assigned *de novo* to these unique features."[4]

Scar never took hold, but *Carolina bay* did, and bays remain as difficult to recognize from ground level today as they were in the 1930s. It's surprising just how many people do not know what a Carolina bay is. For many people it takes aerial photographs to prove just how mysterious bays are.

I have flown over the Carolina bays of Horry County, just as Fairchild's photographers did. Seeing the bays from aloft is an eye-opening experience. As photographs go, altitude and factors like lens focal length determine what we see and the extent of detail. Low-altitude photographs reveal more detail but cover smaller areas. Satellite images cover much larger areas but offer much less detail. Today we can study bays from airplanes equipped not just with cameras but also with LiDAR. It is extremely detailed and precise. LiDAR images, as you see, are colorful.

LiDAR imagery is a relatively new technology. Think of it as a distance technology. An in-flight LiDAR system sends pulses of laser light to the ground. Light reflected back to the sensor, a "return," precisely detects the distance to the Earth. LiDAR generates massive amounts of data. It can send 160,000 pulses per second, creating millions of points that form a detail-rich group of elevation points called a "point cloud." LiDAR helps create digital elevation maps (DEMs) that offer exceptional precision. Laser pulses combined with GPS (global positioning satellite technology) and inertial measurement units (IMUs) measure the roll and pitch of the platform using LiDAR, typically airplanes and helicopters. Working together, the three technologies generate precise, three-dimensional visuals and information about Earth and its surface.

OVERLEAF **A Swamp to Many**
From land Carolina bays appear to be swamps. From the air it is an altogether different perspective, amazing, parallel ovals.

As an airborne laser strikes a target on the ground, the surface reflects the beam of light. A sensor records this reflected light to produce a range or measure. Combining laser ranges with position and orientation data produced by GPS and IMU systems, scan angles, and calibration data, a point cloud results. Each point in the point cloud possesses three-dimensional spatial coordinates (latitude, longitude, and height) that correspond to a particular point on the Earth's surface from which a laser pulse was reflected. Point clouds generate other geospatial products, such as digital elevation models, canopy models, building models, and contours. All this enables scientists and cartographers to examine Carolina bays with accuracy and precision.

From the ground, it's another story. Today, as in 1930, the uninformed that drive by Carolina bays have no idea they are passing Earth's most mysterious landform. Similar to lowland swamps at ground level, people find it easy to overlook the bays. But were they to go aloft high enough, they'd see the big picture—elliptical, parallel, oriented landforms. It's a picture that has generated controversy since the early 1930s, and it shows no sign of abating. As theories conflict and contradict, as proponents discuss and debate, the words of James Dickey in "East of Eden" may have the final say. It's "part of a mystery, part of a whole scheme, a cosmos, the reason for which will never be known."

What is known is that Fairchild's aerial photographs first identified the bays' unique shape and orientation, and people have found them fascinating ever since.

The Great Origin Debate

Origin theories range across the spectrum: beaver ponds, a meteorite bombardment, comet impact, retreating glaciers, shallow ocean currents, spawning fish, ancient buffalo wallows, hydrogen gas "seep" bubbles, melting ice boulders, and dinosaur footprints. Ancient volcano activity makes the list, as do UFOs. Other theories include spring basins, sandbar dams of drowned valleys, depressions dammed by giant sand ripples, submarine scour by eddies, currents, or undertow, and wind blowouts. Still other theories include peat burning by paleo-Indians, basins scooped out by giant turtles, large sinkholes, and solution basins related to artesian springs' activity. Other theories include a secondary impact from a meteor or asteroid impact in the North American ice sheet. The more plausible theories include gradual earth processes. Though less dramatic, the more accepted scenario portrays bays as the product of the combined action of wind and water. It's a theory empirical evidence supports.

References to bay's extraterrestrial origins go way back. The Waccamaw Siouan Indians told a story that thousands of years ago the night sky blazed up as a meteor from the west struck earth. Swamps and rivers filled the depression creating Lake Waccamaw. The North Carolina tribe considered itself the "People of the Falling Star."[1]

Of the hypotheses attending the bays' origin, extraterrestrial theories excite people the most. In 1933 Frank A. Melton and William Schriever first proposed that a mass of meteorites striking Earth created the bays. They based their theory on bays' smooth elliptical shape, sand rims found mainly on the southeastern edge, and predictable northwest-to-southeast orientation.

Geologists tried to corroborate their theory. W. F. Prouty carried out magnetometer surveys in 1952 seeking meteorite material.[2] The surveys proved little. Next he fired .38 caliber bullets into a clay-plasticine mixture to see if their conical shockwaves would form elliptical depressions at a small angle of impact. Elliptical depressions resulted. He concluded a meteorite could produce an elliptical depression two to three kilometers deep with an uneven bottom. Bays, however, are flat-bottomed depressions a few feet below the surface of surrounding land. Most geologists abandoned the bombardment theory. No evidence of meteorite ejecta could be found.

In 1977 Ray Kaczorowski's thesis for his Ph.D. in geology at the University of South Carolina, "The Carolina Bays and Their Relationship to Modern Oriented Lakes," became a book published by the University of South Carolina Press. Kaczorowski disputed the meteorite theory. "Although the development of these depressions is highly variable," he wrote, "absolutely no evidence has been encountered of an extraterrestrial origin."[3]

Five years later, in his *The Mysterious Carolina Bays,*[4] Henry Savage challenged Kaczorowski's conclusions and championed the meteorite bombardment theory. A catastrophic origin like Savage's meteorite theory is sexy. The late attorney's theory had a big problem, though. A meteorite impact would produce shatter cones, shocked crystal rock, meteorite fragments, and iron or nickel elements, crucial evidence that distinguishes extraterrestrial impact structures from terrestrial processes like volcanoes and erosion. Not one meteorite fragment was found in a bay. Science dismissed his theory.

Later a researcher from South Carolina, Christopher R. Moore of the Savannah River Archaeological Research Program, part of the University of South Carolina's Institute of Archaeology and Anthropology, refuted the meteorite hypothesis. "More recent work, including ground penetrating radar, granulometry, optically stimulated luminescence dating, and evidence from examination of LiDAR data, reveals that bays reflect long-term, pervasive and evolving environmental and climatological factors over millennia, not sudden or catastrophic events. Thus, a catastrophic origin is neither supported by geological data, nor needed to explain features we attribute to Carolina bays; Carolina bays are neither enigmatic, nor mysterious, but rather are relatively well understood oriented lakes. While nuances of bay formation through lacustrine shore processes remain to be resolved, the fundamental concepts are well understood and have been for some time. Kaczorowski was right!"[5]

Andy Wood wrote in *Wrightsville Beach Magazine,* "The origin of Carolina bays is more likely owed to long periods of ice age winds during the Pleistocene Period (2,588,000 to 11,700 years before present), when northern North America was covered by an ice sheet more than a mile thick and the Atlantic coastal plain was much colder than present-day conditions." Wood conceded that scientific consensus remains distant, then added, "By most interpretations, persistent winds are the likely cause, sweeping over wide areas of deep sands sometimes covered by shallow water, over time scouring oblong depressions as a result, and eventually becoming wetland and pond ecosystems we know today as the Carolina bay. Think of wind acting on the sand of an open beach, albeit spread over thousands of years."[6]

Dr. Jim Luken, of the Department of Biology at Coastal Carolina University, favors the wind and wave theory. He's seen the process in action. "Up on the coastal plain of the Arctic near the Bering Sea, they have footprint lakes that look just like Carolina bays. Up there it's peaty, and you can watch the ends of the lakes eroding and forming the tips of the ellipses. This less-exciting explanation is probably the better one."[7]

The Comet Theory

Michael Davias of Cintos Research ("Citizen Scientists at Work") proposes an intriguing theory: the "Saginaw Impact Manifold." Davias has used LiDAR images and data to study bays. LiDAR is an excellent tool for mapping and studying bays with precision, but is it a game changer? Davias, who speaks at Geological Society of America meetings and sessions, thinks so.

He speculates that the robustly repetitive Carolina bays may have been generated during a catastrophic mass-transport and deposition of high-purity quartz particles, materialized as a surficial blanket of sand, spread chaotically over an antecedent terrain. "The bays may be imperfections generated within the blanket while the sand was in a state of liquefaction and preserved at lockup as a densely compacted stratum."[8]

Davias advances the idea that an extraterrestrial object indirectly created the bays, a cosmic impact in Michigan's Saginaw Bay area at the time of the Mid-Pleistocene Transition some eight hundred thousand years ago. He theorizes that a "cometary" object hit the Laurentide glacial ice sheet at a shallow angle and hurled ejecta southeast, in a butterfly pattern. "Imagine a snowplow and how it hurls snow to either side," he has remarked. That's how what is now Nebraska and the southeastern coastal plain ended up with Carolina bay–like depressions. (The ice sheet never covered Nebraska or the southeastern coastal plain.) Davias theorizes that the impact ejected superheated ice and sediments outward. As they landed they created shallow basins.

More recently researchers claim extraterrestrial markers such as nanodiamonds and iridium confirm a celestial bombardment theory. Critics scoff, saying these markers rain down every day. Bays' ages also present a problem. Dr. Lee Phillips, associate professor of geology at the University of North Carolina at Pembroke, and other researchers have used core sampling techniques, radioactive carbon dating, and other dating techniques to determine their time of origin. What they found supposedly lays to rest meteor impact or impacts. "The Carolina bays vary rather significantly in their ages, ranging from 130,000 years old to 8,000 years old or less," said Phillips.[9] "They do not date from a single specific time period, meteor shower, or other discrete event. The dates of Carolina bay growth also correspond to times when the earth was relatively warm and free of ice, when wind and lapping Atlantic water could deposit and shape sand on the North Carolina coastal plain." Phillips favors a terrestrial process.

In North Carolina, George Howard, a Carolina bay explorer who co-owns a wetlands restoration business, loves the bays. It's a bit of a legend as to how Howard became so passionate about them. While working as a staffer for U.S. Senator Duncan McLauchlin "Lauch" Faircloth, Howard noticed something strange on a U.S. Geological Survey map of Faircloth's farm. "What are all those elliptical dotted lines on your farm, Senator?" Faircloth said, "Oh, you know, meteor holes."[10]

Howard's curiosity took off, and he studied bays. Howard and friends have dug soils samples from bays and mailed them to an Arizona geophysicist who analyzed them for nanodiamonds, evidence indicative of an extraterrestrial impact. Howard and associates found other extraterrestrial markers in the black mat and in every Carolina bay sampled. Critics, however, say these markers fall to earth every day.

Wind and Wave Action

The prevailing theory is Ray Kaczorowski's "oriented lake genesis." In 1977 research on bay formation and evolution led him to believe Carolina bays evolved from strong, late Pleistocene prevailing winds on shallow, ponded water. Over time the wind created oriented lakes. His NASA-funded report sought to debunk the impact theory by finding the missing piece of the puzzle: where might comparable landforms exist under similar surface influences?

Kaczorowski found what he sought in East Texas, Chile, and on Alaska's North Slope: wind-oriented lakes, he suggested, were bays "in the making." Like Davias, Kaczorowski tried an experiment. He rented a wind machine and set up a model that demonstrated strong prevailing winds could shape natural depressions into ellipses. Furthermore his model oriented bays perpendicular to prevailing winds. Circulation cells created basins and eroded the basins' opposite ends while laying down eroded sediment on the down-wind side. The wind patterns produced sand rims. Kaczorowski referred to this process as "embarrassingly simple."[11] Having explained bays' shape and orientation, he could not explain what force developed the basins themselves. He alluded to topographical irregularities that allow water to pond and then stated that bays can develop where poor drainage exists.

In setting forth his oriented lake genesis, Kaczorowski stated that Carolina bays appear to have been rainfall-fed, or pluvial, lakes. These lakes, with irregular outlines,

Copperhead
The beautiful copperhead discourages many people from
exploring any woods, and Carolina bays especially.

developed during the Wisconsin glaciation and possibly the early Holocene. Impervious organic matter may have enabled ponding. As water levels rose, wind shaped the bays into ovals with long-axis orientation resulting from "end-current erosion" processes. The lakes then developed the "beach/dune" complex now regarded as sand rims, mostly on the southeastern edges, thanks to prevailing southwesterly and westerly to northwesterly winds associated with cold fronts.

But what about the bays within bays so prominent in aerial photographs? Kaczorowski had an answer. A changing climate and dropping water levels produced a succession of beach ridges within the bays, and smaller bays developed in isolated low areas within larger bays. The smaller bays look similar because similar processes that developed larger bays created them.

As for the ages of bays, Kaczorowski said Carolina bays existed as lakes somewhere between six thousand and forty thousand to fifty thousand years ago. He stated, too, the fact that no bays appear to be developing today (in the southeastern United States) probably reflects climactic and/or vegetational change that took place around six thousand years ago. As you see, this does not conform to other theorized ages. Controversy and skepticism endure. If you dig deep into the various theories and their advocates and adversaries, you'll see that things get a bit fractious, chippy even.

The issue of Carolina bays' widely ranging ages remains unresolved, a stumbling block. A single extraterrestrial event would give all bays the same age, and the idea that multiple extraterrestrial events created bays in the same regions is beyond belief. One critic wrote that a bay impact origin is unlikely as the bays were not formed instantly and that evidence indicated the bays were formed before the Younger Dryas period. No one has produced enough evidence to quell the debate, and the mystery lives on. In his thesis Kaczorowski alluded to the fact that many theories on bays' origin were speculation for a quite simple reason. "The bays are generally uninviting, due to extremely dense vegetation, boggy to swampy floors, and the presence of various vermin, including many species of poisonous snakes. For these reasons, it is not at all surprising that most theories are based on intuition, rather than hard field data. Consequently, numerous 'arm chair' theories have been proposed, some very lucid and credible, others bordering on the ridiculous."[12]

Yes, bays can be intimidating, and yes, origin speculation exists. Three things, however, we all can agree on: Carolina bays shelter wildlife, filter and cleanse water, and provide beauty. We need to preserve undisturbed bays and restore as many altered bays as possible.

Dispatches from the Field

It's impossible to escape civilization's essentials, flotsam, and trappings. If I don't see power lines, contrails streak the sky. If I don't see contrails, litter mars highways. Plastic bottles bob along waterways. Windborne grocery bags snag limbs. We blight our habitat and natural places. Even worse, we change wild places' fundamental nature. Carolina bays are no exception. We've drained, timbered, farmed, and destroyed a lot of them. Those that remain pristine, however, are well worth the time to see, for wild places provide a tonic to civilization's ills.

Bays through the Seasons

We visited bays on a seasonal basis to get year-round coverage and document what was taking place. Spring and summer, the seasons of growth, are busy times. We took photos and made notes about water lilies. We recorded the emergence of wildflowers. We noted the changes varying water levels

Pollen Rings
Pollen residue documents the rise and fall of temporary waters.

RIGHT **Lavender Leaves, Golden Pollen**
AIKEN COUNTY, SOUTH CAROLINA
Meadow beauties fleck the bays with color. This small perennial shows up in cypress swamps, moist sands, and boggy meadows. Its stamens' anthers are rich with pollen.

produced. During spring we observed water levels thanks to pollen rings upon cypresses. In spring attention to buttressed trunks revealed white rings.

In the summer we took a close look at the deep pink veins running through meadow beauties. In habitat-specific bays, we got on our knees and hands and searched for Venus flytraps. Examining the oddly beautiful blooms of pitcher plants evoked the mysterious ways of nature. The carnivorous pitcher plant is death made beautiful. We liked seeing white-topped sedge, *Rhynchospora colorata,* in bloom. From a distance you can't see the delicate mint-green tendrils that make up part of the white blossoms. In spring and summer, thanks to morning dew or afternoon rain, we got a kick out of the spiders that build webs across the rims of pitcher plants. The spiders have erected a blockade, Checkpoint Charlie, where they intercept insects.

Toward fall we anticipated sedges' and grasses' fenceless borders where colors define grass zones. Envision a layer of deep blue sky just over a line of green trees and six bands of pink, gold, light green, yellow, dark green, and red grass. Now sprinkle in yellow and blue wildflowers. It's something to behold.

At the edge of sedges and shrubs stand colonies of pitcher plants. Fall will bring colors to them that rival a florist's most imaginative creations. And all these wonderful plants are in close proximity. You can walk from one area of habitat to another in mere feet. Ecotones, areas of transition between biomes, abound in bays. A typical bay gives you the opportunity to explore diverse habitats on foot.

Death's Door
BERKELEY COUNTY, SOUTH CAROLINA
The inventive ways of nature: spiderwebs across pitcher plants' openings capture insects seeking the fragrant "nectar," which will prove to be their undoing even if they escape spiders' ambush.

RIGHT *Still Waters*
BERKELEY COUNTY, SOUTH CAROLINA
Florida Bay mirrors cypress knees and trunks. Beneath still waters life runs deep in a bay. Aquatic insects and amphibians spend part of their life cycle beneath water. This bay has sinkholes, formed from eroding limestone.

Neighboring Habitats

Woods, shrubs, grasses, and pitcher plants form ecotones. In a Carolina bay, a short walk takes you a long way as habitat goes.

As summer fades, signs of fall arrive. Colors begin to change. In autumn sand rims looked as if snow had fallen around and beneath the red leaves of scrub oaks. Even beneath a gray sky, the red leaves of oaks, snow-white sands, and evergreen trees on the horizon painted a pretty picture. In the interior wetlands, we looked for concentrated pools of water where tannic acid looks much like strawberry Kool-Aid. Cypress needles turn orange and stand out against a blue sky.

Sand-Rim Trees

FLORENCE COUNTY, SOUTH CAROLINA

Larger bays feature sand rims on their south-eastern end. These rims show up in aerial photographs. Walking a rim reveals pines and scrub oaks in sands heaped up by long periods of prevailing winds, according to Kaczorowski's oriented wind and wave theory.

RIGHT *A Glimpse of Fall*

AIKEN COUNTY, SOUTH CAROLINA

At Enchantment Bay maple leaves and cypress bark contrast. A peek between the cypress reveals the red leaves of vines and hardwoods.

OVERLEAF *Watermarks*

AIKEN COUNTY, SOUTH CAROLINA

While autumn brightens the bay, cypress bark is darkened where water recently stood. In spring retreating waters leave rings around cypress, the residue of floating pollen.

Up from Black Waters

BAMBERG COUNTY, SOUTH CAROLINA

Cypresses in autumn foliage thrust rust-colored needles into the blue. Out of sight are the fallen needles' mirror-like images upon black waters.

By winter's onset bald cypresses shed their needles, and blackwater better reflects the sky. In the Southeast occasional snows whiten the bays. Leaves of colors float atop the blackwater, creating a kaleidoscope effect as breezes rearrange them. We visited Woods Bay and walked its leaf-strewn boardwalk in winter. We saw fungi on trees and on a warm sunny day found a banded water snake coiled up in a sunstruck patch of needles and leaves.

Just a Photographer and Writer

As we pointed out in our introduction, we are not naturalists or scientists. As a photographer and writer, we appreciate what bays do. They provide much-needed habitat in areas where standing water doesn't exist. Bays provide flood control and make drinking water available to many species. They act as giant, spongelike filters that cleanse water of pollutants and impurities. They provide wildlife habitat and help control erosion. They amaze and unleash wonder. Recognizing that most people will not go into them, aware that many people know nothing about them, we pursue a simple and straightforward mission: we want to share their beauty and wonder and in so doing build support for protecting bays as yet undisturbed.

Going into a bay is not like going into a park. The seasons have a lot to do with what to wear. If it's spring or summer, wear long pants—jeans work—high-top boots, snake chaps as a precaution, and a long-sleeve white linen shirt, which handles the heat well. Wear a wide-brimmed hat or at least a cap. Use sunscreen. Wear good-quality sunshades. Take insect repellent. Mosquitoes abound, as do ticks, and at times deerflies attack in swarms. U.S. Forest Service's Bruce Lyles, retired, says huge clouds of mosquitoes were so thick and ravenous at one bay that he and his colleagues had to run to escape them.

A Wintry Coat

BLADEN COUNTY, NORTH CAROLINA

A rare snow flecks white upon a loblolly pine.
Loblollies make up nearly half of the commercial softwood timber in the South. Clear-
cutting right up to the edge of a bay, however, works a hardship on amphibians and
reptiles during breeding cycles.

RIGHT **Tastes Like Lobster**

Looking like stalactites, this fungus, *Hericium
americanum*, properly prepared, tastes like
lobster, according to wetlands ecologist
Linda Lee.

Palmetto State Icon
BLADEN COUNTY, NORTH CAROLINA
The South Carolina state bird dwells in the high shrubs and forests of Carolina bays. Known for its presence around gardens, sheds, and garages, it also likes pocosins. Here a few snowflakes flutter by an inquisitive Carolina wren in the Tarheel State.

LEFT **Wall of Red**
Besides bringing cooler temperatures, fall light up the bays with color.

Make a record of what you see. Take a compass and a good camera, not a smartphone camera. Always take a lot of water. Don't go alone. Let people know when and where you intend to go. File a flight plan.

Choose your time to go with forethought. Fall is a great time to visit a Carolina bay. Often it is quite pleasant, as the temperatures are cooler and the humidity is lower. Color changes sweep through the bays, giving them a beautiful countenance. Pitcher plants turn brilliant colors, and grasses do too.

In winter take waders and wear warm clothing. Having a few thick towels handy is nice. Again, take a compass and water. Take a good camera, though you won't see as much wildlife. As I said, we are not naturalists or scientists, but like naturalists and scientists, we wade far out into the depths, something you should not do. A rumor we could not run down claims that a photographer went into a bay's watery recesses alone, and his foot slipped between two sunken logs. They snapped shut around his foot. He could not free himself and died of exposure.

Sounds daunting, does it not? Read up on the Carolina bay you wish to visit. Some are so remote as to take themselves out of consideration. Some are off-limits, as the bays at the Savannah River Site are, and because they can't be tampered with, they are some of the more pristine bays around. SRS consists of 310 secured square miles, as it was the site of nuclear weapons development. As the long-time director of the University of Georgia's Ecology Lab at SRS, Whit Gibbons, said, "The best protection for the environment is no people."[1]

A great diversity of life exists at SRS. Skinks, bladderworts, chest-high broom sedge, and wheeling hawks all call this area home. Birdsong is ceaseless. And you can consider it the arena of dragonflies and butterflies.

Wherever you go, whatever the season, be alert and cautious and leave nothing but footprints. If all these preparations and precautions discourage you, if a Carolina bay excursion sounds dangerous, relax in your easy chair and let our dispatches from the field take you inside significant bays throughout the year. To go into a bay is to go into the wilderness. Here, then, is our diary of bay explorations. Let it take you into the wilderness.

Enchantment Bay and Craig's Pond
Savannah River Site
January 27

The day breaks foggy and cold, 31 degrees. Pockets of fog throw our lights back at us, and then we break into clear air only to hit another fog bank. We're on our way to Savannah River Site. We have a bit of a drive if we're to make a 9:00 A.M. appointment with two wetlands ecologists. Both work for the University of Georgia. Linda Lee and Rebecca Sharitz will guide us to some interesting Carolina bays they are studying. We're here to scout several bays for return trips when spring wildflowers burst free of winter and wildlife shakes off dormancy.

The longer we drive, the more downed trees we see. Felled by a recent ice storm, the trees will make the day more of an adventure than we'd like. The fog dissipates.

We arrive on time, and after some involved paperwork, we meet Linda and Rebecca at a prearranged spot. We put our camera gear, supplies, and lunch into Linda's white Explorer. Off we go. Our mutual interest in bays generates plenty of conversation.

We head to Enchantment Bay and its magnificent stand of pond cypresses studded with knees. Linda drives us down a long, straight asphalt road before we turn onto a sandy lane blocked by fallen trees. Maneuvering around downed trees, we manage to keep going. Soon Linda parks, and we all get out. "I have waders in the back," she says.

Into the waders we go. It's a bit awkward walking in waders when you're carrying gear and bundled up for the cold. I feel a bit like an astronaut walking toward the launch pad.

Pollinators' Paradise
AIKEN COUNTY, SOUTH CAROLINA
The Savannah River Site's Enchantment Bay's pond swamp maple leaves contrast with cypress needles' fall foliage.

LEFT *Scenes of Enchantment*
AIKEN COUNTY, SOUTH CAROLINA
At 193 acres Craig's Pond is one of the larger bays at the Savannah River Site. Its vast, grassy savanna hosts invasive maple saplings. As spring greens the bay, flowers splash grasses with color.

Tangles of briars, vines, and thorny shrubs snatch at my hands and clothes. Linda picks up a stout limb and beats openings in the profusion of undergrowth. Soon we see water. We wade into the bay and work our way through more undergrowth and submerged vegetation until we reach an expanse where pond cypresses dominate a grassy, prairie-like stretch of water.

As Robert wades farther out, I take reference photographs. The water around my legs is cold, and my feet sting. We know that come spring we'll be back photographing and documenting the plant and animals in this rich repository of life known as Enchantment Bay. As we wade back, Linda pulls up a carnivorous plant, a bladderwort. Underwater it opens up; in dry air it looks a bit like a thick strand of algae. It's so green it's fluorescent.

Back in the Explorer, we head out and make our way to another sandy lane. The object is to see one of the larger bays in the area. Fallen trees and ice storm debris block the sandy lane leading to Craig's Pond. We get out on foot and clear trees so Linda can drive her white Ford Explorer to the bay. Craig's Pond is impressive, one of the larger

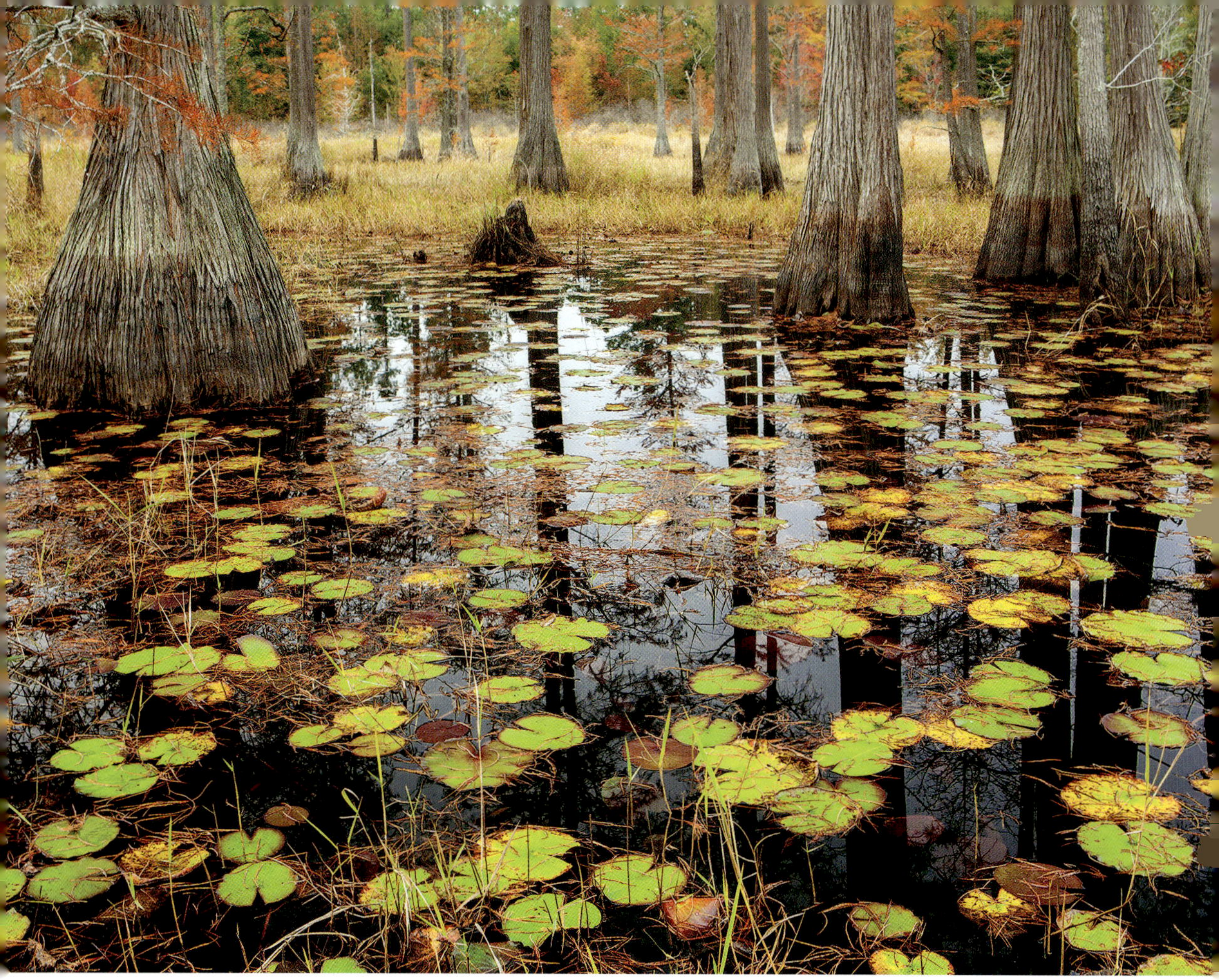

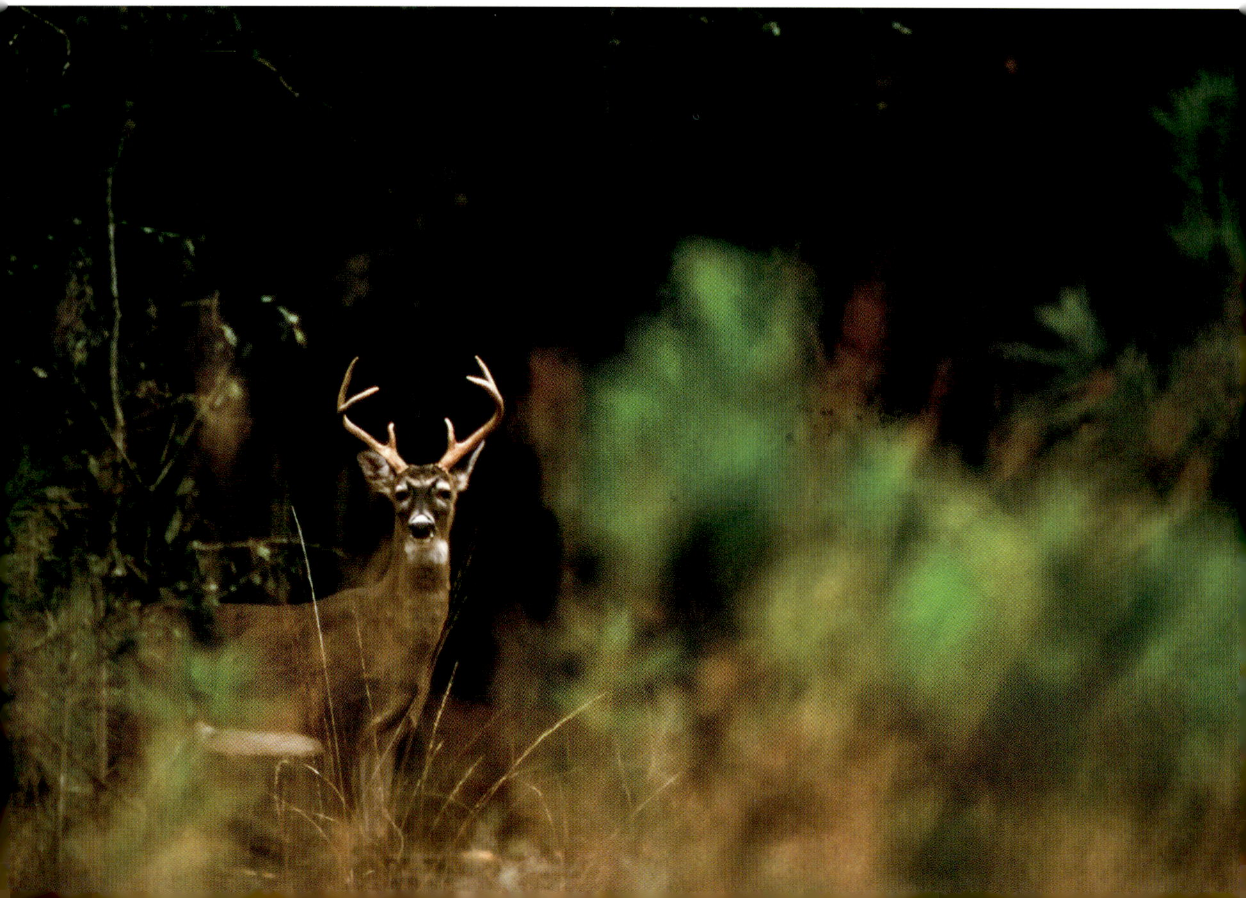

bays at SRS. We see deer. About eight or nine large wild hogs are rooting around.

Downed trees make the going rough, and finally we reach a stretch of road that's impassable. We get out and walk. The largest deer I've ever seen bounds through the woods. It leaps in great, graceful arcs, its white tail flashing.

A long walk brings us to the sandy edge of a bay. At 143 acres it's fairly big. Walking into its grassy plain, we see trails tunneling through the grass. Some are deer trails; it looks as if massive anacondas slid through. We see hog wallows too, and then we spot more wild hogs. As Robert stalks the hogs, a song comes across the plain: frogs serenade us. By now it's noon, and the temperature is in the low fifties.

You have to be here to appreciate the vastness of Craig's Pond. Rebecca points across its vast grassy plain. "I've canoed across the entire area," she says. The far tree line bristles with topless trees, victims of ice. This fact speaks to the rainy season and how the bay fills with water. We will return when spring sets in, dragonfly time.

Back in the Explorer, we scout Sarracenia and Dry Bay and ride through the remnants of Ellenton, where we scout a bay with a high-tension powerline running through it. Locals grew asparagus in this bay. In the early twentieth century, it was discovered that local conditions proved perfect for asparagus, and until the early 1930s more asparagus was grown and shipped from the Williston area than from anywhere in the country. Williston became known as "The Asparagus Capitol of the World."[2] (California gradually captured the market.)

We make our way to a main road, and after passing through an inoperative checkpoint, we soon pass a monolithic building. It's R Reactor, huge and foreboding. "You aren't allowed to photograph that," says Linda. (When security badged us in they warned us not to photograph buildings.) The baleful building we can't photograph is the site's first production reactor. It once produced tritium and plutonium-238 and -239.

For half a century it's sat idle. In 1963, after ten years of operation, a defective fuel rod released cesium-137 into Par Pond, a large lake built to provide cooling water to two reactors. President Johnson called for a reduction in the arms race in January 1964. Six months later R Reactor closed. The day was June 17, 1964. Just looking at the reactor amounts to a world history lesson. For all we know, that building may have helped stave off global calamity.

In 2011 the decommissioned reactor was filled with low-density cement injected with foaming agents. That turned the reactor into a solid, impenetrable block. Its doors were welded shut. It's expected to stay this way for the next 1,400 years, an astounding amount of time to us, a second to the universe. And yet just down the road are natural areas like no others. I refer to SRS as an "atomic paradise" because the natural world there is more natural than most other places, save wildlife refuges and national parks, and in some ways it trumps those. In one of this country's great paradoxes, establishing a facility to refine nuclear materials produced one of the more protected natural areas in the country. How many Walmarts, parking lots, sewer systems, and general development would have plagued these 310 square miles had there been no SRS.

Here you find solitude. The wind whispers through pines, and you cannot hear one sound from civilization. The sky is free of contrails. No litter. No billboards. No traffic. It's like going back three hundred years, and it should stay this way, since Craig's Pond and Sarracenia Bay are in the Department of Energy's Set-Aside Program, a system of reserve areas established on DOE facilities to provide reference sites for understanding human impacts on the environment.

Woods Bay State Park and Heritage Preserve
Latitude 33.9486, Longitude −79.9783
Olanta, South Carolina
May 1

A cloud cover diffuses light. It's a good day for photographing Woods Bay and its abundance of wild things. Yesterday's rain of two inches has the water high and black, and insects dance, skate, and scoot over the water like Central Park ice skaters.

We're here because Woods Bay is one of the Mid-Atlantic coastal plain's few large, intact bays left. It's unusual in that it's spring fed. Spilling into Florence, Clarendon, and Sumter Counties, Woods Bay includes marsh, sandhills, oak-hickory forest, and shrub

bog. It's easily accessible, being just south of Interstate 95. In the 1920s most of the cypress trees were logged. In 1971 logging companies were prevented from cutting there again. Woods Bay was made a state park and, later, a heritage preserve. In it you'll find swamps, savanna areas, and a pine barrens along its sand rim. A cypress-tupelo community dominates its interior. You'll see wildflowers here.

Consider it "easy man's bay." Easy to get to, easy to see, and easy to explore by walking the 1,150-foot boardwalk that takes you through the dense, impenetrable vegetation. You can drive right by a part of its white sand rim. A bonus is the millpond joining the bay that, though heavily grown up, affords views of wildlife from the nature trail encircling it. It's a place where you can see alligators. More than 75 species of mammals, reptiles, and amphibians and more than 150 species of birds, some seasonal, live here.

In Woods Bay the trills of frogs and repetitious cries of woodpeckers accompany a chorus of songbirds. (Spiders play their deadly waiting game in silence.) An undisturbed bay is not quiet. It's filled with creatures' cries and calls. Somewhere among the buttressed cypresses, a frog sounds like someone chopping wood. What seems to be a large frog, judging by the volume of its call, sounds like a stack of planks dropping to the ground. A bird looking much like a robin was not, for its call sounds like that of a rainforest bird in the Amazon or Congo Basin. A flash of gold disappears into the pocosin. Among leaves and humus hide salamanders. Prothonotary warblers? They're here. You won't hear wildlife sounds or see warblers in adjacent, destroyed Dials Bay, Woods Bay's twin.

Spotted Salamander

Spotted salamanders live in woodsy habitats close to vernal pools, swamps, and stagnant waters. They burrow into damp surroundings and find habitats with temporary water safe as no fish can prey on its eggs or larvae.

Caterpillar Attack

In twenty minutes this band of caterpillars stripped vegetation to its stems.

The wind coos through the pine tops, and birdsong fills the air. Frogs override the buzz of flying insects. Some frogs do indeed croak. Others bark, and some boom.

Our footsteps betray our presence. As we approach, frogs leap into the water. Startled skinks hurdle into the water, where they lift their heads and skate across the blackwater like surfers. One skink, lime green with a copper head, stands apart from his electric-blue peers. Creatures from below pucker the water. Raucous blue jays pierce the air.

The pure green of spring leaves and cypress needles shocks the eye. It's hard to believe such freshness exists. Rampaging caterpillars devour the new, tender leaves. In the span of twenty minutes, they shred leaves, leaving naked stems. In droves they undulate. On signal all undulate in unison, causing the host plant to oscillate. Each caterpillar is black with yellow fuzz along its sides. Eleven pairs of yellow dots mark the length of their backs.

Woodpeckers continue to hammer trees, crying out between blows. A great blue heron soars overhead, his sore-throat cry serving notice he's there.

No gators about—that I can see—but they're here. A curious observation. No squirrels at all. The trees here stand in and over water. Is that reason enough for squirrels to stay away? One fall and they become gator bait?

Methane bubbles pockmark the black water, and over from the direction of the mill-pond comes a roar akin to a diesel truck coming to life. A bull gator. I've transported myself to Joseph Conrad's Heart of Darkness here at easy man's bay.

Lewis Ocean Bay Heritage Preserve
Latitude 33.795, Longitude -78.848
Horry County, South Carolina
May 22

We're here to see Venus flytraps. Our excursion into this bay was unusual in that we had a guide. James Luken, Ph.D., biologist and associate provost at Coastal Carolina University, led us into this bay known for its Venus flytraps. Luken studies wetland ecology and carnivorous and aquatic plants.

Lewis Ocean Bay Heritage Preserve is an important bird area. At least twenty Carolina bays exist on this 10,097-acre preserve. Like many bays, Lewis Ocean Bay exhibits seasonal splendor. January brings wiregrass and migrating songbirds. March through June, pitcher plants and wildflowers are dazzling. In the fall wildflowers such as meadow beauties pop up.

Here you can see Venus flytraps, yellow pitcher plants, lichens, native orchids, black bears, bald eagles, and red-cockaded woodpeckers. Carolina bays' habitat diversity suits Luken's interests to a tee. "What makes bays unique are the different habitats, wild plants, and wildlife," says Luken. He refers to the bay we stand in, Lewis Ocean Bay. "There are a dozen or so habitats in this preserve. The interior parts are thickets of evergreen shrubs that create great habitat for black bears and bobcats. In fact it's prime habitat for black bears. They spend the winter in the dense vegetation and seek refuge there in summer."[3]

10,097 Acres, 20 Carolina Bays
This complex of bays provides habitat for black bears, Venus flytraps, and pitcher plants and is a natural area important to migratory songbirds.

Lewis Ocean Bay's varying water depths, sand rims, and peat mats nurture botanical stars such as water lilies, sedges, red bays, sweet bays, butterworts, sundews, gallberry, and white- and yellow-fringed and rosebud orchids. Add the rare Wells pixie moss too.

"Around the edges where you move from low, wet, thick vegetation to the dryer, not-so-thick vegetation, there's a little transition zone ten to twenty yards wide," says Luken. "That's where you find a lot of rare plants." Plants like the Venus flytrap, which Charles Darwin called "one of the most wonderful [plants] in the world."[4]

Venus flytraps grow where sandy, nutrient-poor soil changes from wet to dry beneath showers of sunlight. Unwary insects provide what poor soils cannot—sustenance.

FAR LEFT **Two Life-Forms in One**
HORRY COUNTY, SOUTH CAROLINA
A bay's sand rim hosts lichens, green, sponge-like plants that grow in clumps. An other-worldly, complex life-form, lichens comprise a symbiotic partnership between a fungus and an alga. The fungus gives the lichen most of its characteristics.

Green Barricade
HORRY COUNTY, SOUTH CAROLINA
An example of a pocosin at Lewis Ocean Bay. While *pocosin* refers to a type of wetland with deep, sandy, acidic, peat soils, it also references an evergreen shrub bog community growing in often-wet, nutrient-poor soils, as in bays.

The flesh-dining foliage finds that any insect will do, as will humans—well, in Hollywood's imagination, at least. *The Little Shop of Horrors* (1960) crossbred a Venus flytrap with a butterwort to create a plant that ate humans. In *Venus Flytrap* (1970) a demented scientist uses thunder and lightning to turn carnivorous plants into man-eaters.

It's illegal to take Venus flytraps from their wild habitat and unnecessary. Legal avenues exist for buying them. The North Carolina Department of Agriculture's endangered

The Legendary Venus Flytrap
HORRY COUNTY, SOUTH CAROLINA
Charles Darwin called the Venus flytrap "one of the most wonderful [plants] in the world." It's found only in Carolina bays and pine flatlands within a ninety-mile radius of Wilmington, North Carolina. How it kills: A nectar-like scent attracts insects to its open, specialized leaf and its six sensitive trigger hairs. When an insect bumps the first trigger hair, nothing happens. Touch the second trigger hair, however, and the leaves snap shut in twenty seconds. Enzymes then digest the doomed insect.

species division licenses some growers to harvest seeds and grow and sell flytraps. People who knowingly buy these Hollywood-inspiring stars from poachers, however, contribute to a black market, and thus poachers head to Venus flytrap habitat. "Once you see the environment where they grow, you instantly become the expert," says Luken. "You can eliminate 90 percent of the rest of the area and go exactly where they're going to be."[5]

We looked for the wondrous plant, a bucket list item, to photograph only. It wasn't easy to find. The leaves are about the size of your thumbnail. After coming up empty for a long time, I found several dozen flytraps, all beautiful, some about to bloom, and all with a leaf or two shut, meaning the plant was dining on some hapless critter.

After an hour of bumbling around, I got better at finding them, a scenario, sadly, that poachers know all too well. Once these thieves learn where flytraps grow, they find them with ease, dig them up, and sell them.

Less than thirty-five thousand remain in the wild, and it could be that far fewer than that live in native habitat. Another problem is loss of habitat to development. This mysterious plant is creeping toward extinction. Up in North Carolina, the number of counties growing flytraps has fallen from twenty to twelve. South Carolina has them too. They grow in a narrow band about ninety miles long from South Carolina into North Carolina.

So how does the Venus flytrap spring its trap? Its leaves ooze a sweet nectar that attracts insects. They come to the leaves to check things out. In the leaf wait six sensitive trigger hairs. A fly lands on a leaf and moves about. Touching the first trigger hair does nothing. As soon as it touches the second trigger hair, however, a timer counts down. In twenty seconds the leaves snap shut, their outer tips locking together like prison bars. The battle is over, and a ten-day digestive period then takes place. It's one of nature's more unusual clashes.

Signs of human conflicts exist here, too. When we first set foot in the bay, Luken told of finding machine-gun shell casings from Air Force jets that strafed the area during practice runs. He said they were getting harder to find. The late novelist James Salter was stationed at Myrtle Beach, and he wrote about how wastelands made good gunnery ranges for the base's pilots. I looked down at my feet, and there lay two casings as big as thumbs and as long as index fingers.

Luken went to a meeting and left us in the bay. Leaving, we came upon a yellowbelly turtle crossing a sandy stretch. Later, in a phone call, Luken asked me if we saw any bears. "No, I said. "Well you should have," he said. "About three hundred bears are in Lewis Ocean Bay."

Bay Bound

HORRY COUNTY, SOUTH CAROLINA

A bit aloof, a yellow-bellied slider, *Trachemys scripta*, plods toward a bay at Lewis Ocean Bay Heritage Preserve. Slider turtles, abundant in ponds and streams of the Southeast, live in habitats that include Carolina bays, sloughs, sinkholes, oxbow lakes, swamps, rivers, lakes, and ponds. They occasionally travel between bodies of water.

LEFT **The Symmetry of Nature**

CHARLESTON COUNTY, SOUTH CAROLINA

You'll find male and female reproductive cones on the same longleaf pine. Despite having "wings," cones' large, heavy seeds don't fall far from the tree. Mice, birds, and squirrels feast on the seeds.

A Rare Daytime Appearance

HORRY COUNTY, SOUTH CAROLINA

Big and stocky barred owls cruise the bays
come sundown seeking prey.

Cartwheel Bay
Latitude 34.1462899, Longitude -79.0797981
Horry County, South Carolina
May 26

The 568-acre Cartwheel Bay Heritage Preserve in Horry County protects one of the few known Carolina bay longleaf pine savanna complexes in South Carolina. The savanna here is overgrown with shrubby undergrowth, proof that no fire has passed through of late. Lichens appear here and there. The wind, rising and falling in the pine tops, sounds like surf. It's hot but pleasant. We wear snake chaps, but so far all we've encountered are small frogs leaping into water.

Two bays here sit close together: Mossy Bay and Cartwheel Bay. A lonely bobwhite whistles, and a barred owl answers. A deer leaps through undergrowth, white tail erect. Sporting its reddish summer coat, it vanishes into greenery. Crows and red-cockaded woodpeckers put up a ruckus. Then it's quiet again. Nothing but wind in the treetops.

Waiting for Prey
RAINBOW BAY, AIKEN COUNTY, SOUTH CAROLINA
A common species, the American green tree frog proves popular as a pet. It's best suited for the wilderness, where it's truly at home among lily pads, vegetation, and watery haunts. Flies, crickets, mosquitoes, and other small insects make up its diet.

Rainbow Bay
2.4 Acres
Savannah River Site
May 29

On the long road to Rainbow Bay, we see a reminder of a childhood story, three little pigs. Two are red, and one's coat is black-and-white polka dots. They scurry into high grass, gone in the blink of an eye. They're cute but will become a menace—wild hogs that maul habitat.

Rainbow Bay, an isolated seasonal wetland, sits in the center of the Savannah River Site. Within an 87.5-acre set-aside area are Rainbow Bay, a forested buffer area, and a wedge-shaped corridor extending to a tributary of Fourmile Creek. The corridor

Where Wildlife Drops In

Unused fuel rods hold up flashing, and a drift fence results. Researchers bury buckets flush with the ground along it at regular intervals and check them daily. Amphibians and other animals fall into the bucket and await their rescuers, who make records of what they find.

provides a woodsy conduit between Rainbow Bay, Bullfrog and Pickerel Ponds, and the Fourmile Creek watershed. This corridor lets amphibians move freely and relatively safely among these wetlands, a path to water and water lilies. Before the Savannah River Site was built, farm fields surrounded Rainbow Bay, again proving that the construction of SRS was a blessing to its Carolina bays, and a unique record resulted from it.

Rainbow Bay's amphibian reserve earned its way into the *Guinness Book of World Records* for its ongoing sampling of the amphibian and reptile communities. The study began in September 1978. Daily animal population censuses make this project the longest-running community study of its kind in the world. Researchers installed a drift fence around Rainbow Bay, a sampling method whereby a fence of flashing (like roofers use) is erected and forty-four buckets placed in the ground on each side of the fence at regular intervals. During the night amphibians and other creatures travel along the fence until they fall into a bucket. Researchers visit the fence each morning and count the amphibian and reptiles. They identify, measure, and mark them prior to release. It can get busy. Hundreds of salamanders or thousands of baby spadefoot toads have been found in just one bucket.

This ongoing census, a model for long-term studies, helps us understand the effects of climatic fluctuations and habitat disturbances on natural amphibian populations. That Rainbow Bay Guinness World Records certificate encased in plastic? Sunlight has faded the words, and water stains did their part to make the official proclamation difficult to read. You'll see other signs at SRS as well—yellow-and-black metal signs: "No Trespassing by Order of the United States Department of Energy."

The upshot is a one-thousand- or five-thousand-dollar fine, imprisonment for a year, or both if you come onto the property with weapons and explosives. So the public can't access Rainbow Bay at the Savannah River Site, but the bay has a long history of interaction with man, thirteen thousand years. Ancient dwellers found sustenance here, and so did I. As we walk past a drift fence, Linda Lee snaps off a delicate tendril of smilax. "Here, eat this. It tastes like a fresh bamboo shoot." I do, and it does.

We make a quick stop at Enchantment Bay. Since we were here in late January the sweet gums have taken over, a sign of dry weather. Linda uses a machete to hack through the sweet gums and briars that close off access to the bay. Getting into some bays is no easy task.

Craig's Pond–Sarracenia Bay, Mona Bay, Enchantment Bay, and Ditch Pond
Savannah River Site
May 29

Today is a busy bay day. We visit or revisit four bays at SRS—Craig's Pond, Sarracenia Bay, Mona Bay, and Enchantment Bay—and on our way back a fifth near Williston, Ditch Pond, which spreads across Aiken and Barnwell Counties.

Craig's Pond and Sarracenia Bay—People like to talk about how peaceful and quiet nature is, but that just isn't true. In Craig's Pond the cry of hawks and the rustle of skinks scooting into cover is constant. Doves coo. As in other bays, the birdsong is ceaseless. When you stand in Craig's Pond's savanna, birdsong is a 360-degree experience. Chest-high broom sedge carpets the bay. Bladderworts bloom. It's the arena of butterflies and big orange dragonflies that look like biplanes. Predation is ongoing.

Only the faraway whine of a commercial jet blemishes serenity. Humanity, with its propensity for delineating its property, has run a fence straight through the chest-high sedge. Were it not for that airplane and fence, you'd think humankind doesn't exist in this wild place, but wild hogs clearly do. Their wallows make a mess. Even so, among the hogs' rooting bloom purple bladderworts. Two hundred yards away, big, brown wild hogs forage in the sedges.

Gray Skeins, Airy Draperies
AIKEN COUNTY, SOUTH CAROLINA

Though far from the Lowcountry, the trees at Ditch Pond bear heavy loads of Spanish moss. Of course the plant is neither Spanish nor moss. It is a bromeliad and in the same taxonomic family as pineapples. Most bromeliads, including Spanish moss, are epiphytes, which grow on other plants but do not rely on them for nutrients. They take nutrients and moisture from the air.

Craig's Pond, at 193 acres, is one of the largest true bays at SRS. Sixty-four percent of the pond, however, is off-site. Sarracenia Bay lies to the west of Craig's Pond and is 9.9 acres. In 1993 and 1996, both bays provided sustenance for the foraging wood stork, once incorrectly called wood ibis (it is not an ibis). The wood stork is the only stork that breeds in the United States.

Mona Bay—This bay, recently burned, sports maiden cane, which grows in thick

Ditch Pond's Golden Waters
Bladderworts bring lemon tones to the shallow waters of Ditch Pond, a "high pond" lush with aquatic plants.

mats that cover the ground. Walking across it feels like walking over a web of dense, coarse fabric. Mona Bay, not quite twenty-eight acres, contains areas of open water with herbaceous and aquatic plants. No frog choruses. Apparently it's not the time of day for frogs in this depression wetland.

Enchantment Bay—A storm blows in and washes out the tour. Thunder rumbles and the rain falls as the devil beats his wife.

Ditch Pond Heritage Preserve—Daylight fades fast as we pull into the parking area for Ditch Pond, a twenty-five-acre bay. As I walk over to the exhibit that orients visitors,

BERKELEY COUNTY, SOUTH CAROLINA
Multicolored Haze
Summer sees its share of wildflowers. At Red Bluff Bay, if you sit low to the ground and gaze across the tops of sedges and grasses, it's like peering through fog with bright lights breaking through.

RIGHT *Pitcher Plants Overlook the Savanna*
Pitcher plants edge zones of sedges and grasses that color Red Bluff Bay's savanna in the Francis Marion National Forest.

I find a crumpled five-dollar bill on the ground. It pays to visit bays. We get our cameras and head into the bay. Right away the abundant, thick draperies of Spanish moss are stunning. The trees bear heavy loads of moss, the most I've seen anywhere. It gives the site a beautiful yet funereal air. The ruins of a cemetery reinforce that impression.

Referred to in the past as a "high pond," lime sink, and sink pond, Ditch Pond, wide and flat, is lush with aquatic plants. Water lilies abound, and walking out on its boardwalks affords good views of the aquatic plants. Eight rare plant species of concern exist at the preserve: blue maiden cane, Robbins' spike rush, creeping St. John's wort, piedmont water milfoil, awned meadow beauty, slender arrowhead, Florida bladderwort, and piedmont bladderwort.

Red Bluff Bay
264.25 Acres
Francis Marion National Forest,
South Carolina
June 15

Red Bluff Bay's flowering plants and pitcher plants put on quite a show. For now the show is delayed. At 4:30 in the afternoon, a string of storms keeps us in the SUV for thirty minutes. Standing in a flat bay next to a tripod is not a good idea. We wait out the storms that help provide the bays' lifeblood. As the lightning moves out to the southeast, we head into the bay's interior. Soon we stand in a pitcher plant bog among colonies of green and yellow pitcher plants and hooded pitcher plants that resemble a doe—soft white spots against a tan background. Blue irises and white top sedge fleck the grasses with purple and white, and if you look closely, a delicate green attends the white top sedge blooms. Gaze across the tops of the sedges and grasses, and it's as if you peer through a layer of fog. Here and there trees stand amid the grasses.

A Florist's Delight
BERKELEY COUNTY, SOUTH CAROLINA
Florida Bay looks like a flower garden come spring. Just feet away stand pitcher plants. Scan the bay's interior, and colors pop out of the surrounding greenery.

Frogs, fading peals of thunder, and birdsong fill the air. The coo of a mourning dove and the dulcet call of a bobwhite would delight sportsmen and -women. Every bay, it seems, has its resident bobwhite, but doves are more common. Small frogs impossible to see give out a metallic clicking song. It sounds much like someone running a thumb down the length of a plastic comb. Some frogs whistle, and some whimper like puppies.

At the sedges' edge, brown birds a foot apart dance a duet in the air, like two French Mirages duking it out. Each mirrors the other's moves.

Bays surprise you. What you see and what you don't see prove significant. You can stand in the savanna and trace the wood line until your eyes complete an elliptical path. The compass reveals that the bay's axis runs northwest to southeast, and I see an imaginary line dissecting the bay, but I don't see power lines, concrete, gum wrappers, or plastic bottles, and for that I am grateful.

Florida Bay (Wambaw Bay)
50.53 Acres
Francis Marion National Forest, South Carolina
June 16

We're here to see what's possibly the best Carolina bay to be found. We come, too, to see how Florida Bay's wildflowers fare. Just as we arrive, a turkey hen and her poults run from the bay right past us. Other than deer, wild hogs, and alligators, they're the largest creatures we've seen. Here and there, wildflowers such as meadow beauties and white top sedge color the bay, but the pitcher plants dominate. Insects fly all around them. Without doubt the star attractions today are pitcher plants, so named because their throat is like a long, lean pitcher. Insects catch the fragrance of the nectar and fly

or crawl into the long throat. Escape is difficult. Downward-pointing hairs block the way out, and the insect becomes the victim of the very nectar it hoped to sample. Falling further, it ends up in a broth of juices similar to stomach acid. There it turns into food for the plant.

The pitcher plants are vibrant. Unlike in the fall, when they rattle in the wind, now they are succulent. The fragrance of their nectar pulls hard on insects. Throughout the colony flies, dragonflies, spiders, and wasps land on them. The plants are bug magnets. Spiders have learned to stalk pitcher plants, to spin webs across the tops of the pitchers to steal insects from the plant. It's an ingenious strategy and shows how efficient nature is at evolving survival tactics. And it's beautiful. Morning dew makes the webs sparkle.

Dining in Dappled Light
CHARLESTON COUNTY, SOUTH CAROLINA
Startle this bay-browsing white-tailed deer, and you'll see how it gets its name. The tail will fly up, revealing its white underside as this herbivore flees.

BELOW **Sarracenia flava**
BERKELEY COUNTY, SOUTH CAROLINA
The sweet or yellow pitcher plant wears its summer coat. When autumn comes and colors fade, winds rattle the dry, Styrofoam-like plants. A pitcher plant is a modified leaf, not a flower.

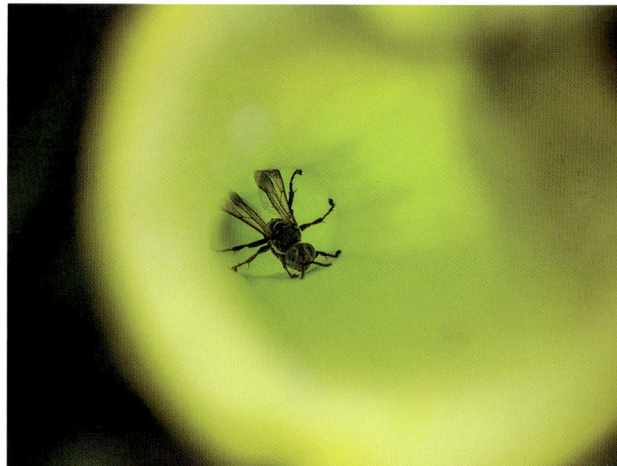

CLOCKWISE FROM ABOVE

Lying in Wait

BERKELEY COUNTY, SOUTH CAROLINA

A patient nursery web spider waits and waits. The pitcher plant will lure prey, and the spider will be ready. At times like this, the plant's fragrant "nectar" presents a dual threat. Insects beware.

Trapped

This wasp got lucky. The plant didn't turn him into its own brand of soil.

Another Deception

BERKELEY COUNTY, SOUTH CAROLINA

Some hooded pitcher plants' feature soft white spots against tan, resembling a doe's coloration. The white spots look like openings to insects, but when they repeatedly try to exit through them, they burn valuable energy that hastens their fall into the acid.

As I make notes, the sounds of an epic struggle drift up from a colony of pitcher plants. An insect is fighting for its life. A frantic buzzing comes from a cluster near my camera. Carefully I lift the lid of a plant. Nothing there. I try another. Nothing there. I lean over and listen closely, then lift the lid of a third plant, and inside a trapped wasp fights for its life. Its wings are a blur. The front legs claw at the plant's leaf tube. Slowly it rises to the top, only to fall deeper than where it started. Now the wings beat with fury, and the buzzing amplifies in the trumpet-like tube. Even though wasps stung me in the face twice in this bay, it's tempting to rescue it, but nature must run its course. I step away but return a few minutes later. All is quiet. I lift the lid, and the wasp is quiet. Suddenly it beats its wings wildly and rises to the top, only to fall again.

This struggling dooms insects. Eventually they tire and sink lower and lower until the downward-pointing hairs make escape impossible. The finale in this life-and-death drama is a plunge into the digestive enzymes where insects dissolve and provide the plant nutrition.

Some hooded pitcher plants, the ones spotted like a doe, spring another trap on insects. The white spots look like windows or openings, and insects try to exit them only to expend valuable energy that hastens their doom. Meanwhile our valiant wasp continues its fight for freedom. It has a factor in its favor—size. Is it too large to fall among the downward-pointing hairs?

The wasp struggled for twenty minutes. Then all was quiet. I stood a few feet away and listened. Nothing. I lifted the lid expecting to see it awash in the plant's acid, but the panic-stricken wasp buzzed with renewed ferocity, and a miraculous thing occurred. It flew out, a rare escape from a plant evolution perfected into a deathtrap. Apparently the wasp's size kept it from sliding into the hairs that act like the crossbars of a jail cell.

Thunder Bay
Savannah River Site
June 20

Thunder Bay is small, with deeper water near its center where water lilies grow thick. Linda Lee praises water lilies as an almost religious icon. "The Eastern lotus, a completely different genus but shares some traits, has long been a religious symbol in the East." She drew the parallel with our fragrant native water lily when she was a botany student trying to collect one for her required collection of pressed plants.

"I came across a borrow pit full of the most disgusting-looking, murky, green, probably eutrophic water I've seen and some fragrant water lilies in bloom. It was deep, but I waded in to get one, grasping the flower-stalk below the water. When I tried to pick

Water Lily Nirvana
Wetlands ecologist Linda Lee is at home among Enchantment Bay's water lilies. She sees a connection with the waxy, water-repelling water lily bloom and the Eastern lotus, or *padma*. "This beautiful, unspoiled thing that thrives in the muck and grime, yet is never tainted by it."

ABOVE RIGHT **The Symmetry of Perfection**
AIKEN COUNTY, SOUTH CAROLINA
A beautiful, beneficial water lily opens in a Savannah River Site bay. The pure-white petals shed water with ease and shade fish and the pads provide places of rest. A flotilla of snow-white blooms upon black bay waters is not easily forgotten.

it, it drew the open flower under the murky water, but to my surprise, it re-emerged just as clean and fragrant as ever. Hard not to see a metaphor there, no? This beautiful, unspoiled thing that thrives in the muck and grime, yet is never tainted by it. Eastern lotus is even more water repellent than water lily, a nanosurface famously studied by engineers. I have no real knowledge of why the eastern water lotus became a symbol. It may have nothing to do with this. But it is easy to see why it would captivate any mind seeking a higher plane of existence."[6]

As I consider Linda's observations, I realize SRS itself is as mysterious as the bays are. Today we drove by P Reactor, a strangely beautiful gothic cement monolith. It never gets old to think about this place's duality: nuclear bombs and radiation in the Garden of Eden—an atomic paradise filled with wildlife. Today we've seen deer and turkeys. No snakes.

Large flies strafe me, their buzzing rising on approach and falling as they zoom away—a Doppler shift. The flies crash into the UGA Ecology Lab's white Ford Explorer

#91421 that's in their flyway. A woodpecker laughs like a hyena. Then it's summoned. Never to be heard again.

It's 12:55 P.M. at Thunder Bay, and the cricket frogs sing their amphibian refrain. Jackal-like cries from woodpeckers respond: from the northeast one sounds, and from the northwest another answers. Describing all this birdsong proves challenging. Imagine chatter, squeaks, whistles, peeps, warbles, scratchy notes, and pure calls such as the bobwhite quail and the guttural, almost coughing of crows and now and then the piercing cry of hawks. Now mix it all together in a great soundtrack and pipe it from speakers surrounding you in a perfect circle. That's what it's like to stand in Thunder Bay on a June afternoon. Just when you think it can get no more majestic, the rising and falling music of cicadas joins the mix. A blue jay sounds off, doing his best to mimic an impatient driver honking his horn in congested traffic. And then far away comes the rumble of thunder. Thunder Bay lives up to its name.

Dalzell Bay
Sumter County, South Carolina
July 22

We're at Dalzell Bay, so named by Steve Bennett and John Nelson for the town nearby, back when they researched their pioneering 1991 *Distribution and Status of Carolina Bays in South Carolina,* a study heralded today.

It's a reunion. Steve, Robert, and I are in the field together for the first time since 1987, and we're meeting the Saginaw Impact Manifold origins theorist Michael Davias at 7:30 A.M. at Cannery Road near the bay. Davias has driven all night from Stamford, Connecticut. We're eager to discuss the

Imminent Takeoff
SUMTER COUNTY, SOUTH CAROLINA
At Dalzell Bay two biplane-like dragonflies take a break from aerial acrobatics. As dragonflies dart about seizing prey, the buzzing clatter of wings fills the air spring and summer.

bays, their wealth of habitat, uniqueness, and origin theories. It won't take long, however, for this bay to remind us that bays can be inhospitable places.

After meeting at Cannery Road, we enter woods full of tangles and logs and make our way west to the bay's pond cypress swamp, for now out of sight. The woods are thick, and we use a compass and satellite imagery to make our way to the bay's mirror-like black water. Here and there remnants of garbage piles mar the view—rusted-out cans and spent baby food jars. Shotgun shells litter the ground, and we come across

Dalzell Bay
SUMTER COUNTY, SOUTH CAROLINA
The contrasting beauty of soft grasses and cypress.

a tree stand. Besides the litter, nearby Shaw Air Force Base sends jets up every few minutes. "I feel like I'm at LaGuardia," says Davias. The roar is deafening. Gnats fill the air, another nuisance. Dragonflies don't have to worry about food here.

Not long after we set foot in the bay, Robert plunges into a stump hole and goes all the way under, ruining a camera and cell phone. This mishap cuts our day short. Back at my car, locals have convened to see if all is okay. We're warned about all the moccasins at the "duck pond." Someone should have warned us about stump holes. This expedition? Our most disappointing.

An autumn trip to Bladen County, North Carolina's Horseshoe Lake, a six-hundred-acre Carolina bay, was more rewarding. Grasses grew up to the base of cypresses, creating a setting of contrast in colors and life-forms: soft, golden grasses and hard, gray-brown cypress. Yellow pitcher plants grow there, but the autumn grasses and cypresses offer uncommon beauty.

A Bay in the Making?
AIKEN COUNTY, SOUTH CAROLINA
Aiken County's Janet Harrison High Pond is a distinct depression on fairly flat, well-drained upland ridges. Some refer to high ponds as "depression meadows." This depression may be one of the stages leading to the formation of a Carolina bay. The problem is that no one knows for certain how bays form.

Janet Harrison High Pond Heritage Preserve
16.85 Acres
Aiken County, South Carolina
July 23

We're in peach country to see a Carolina bay known as a "high pond," a common term used for bays in South Carolina's northwestern coastal plain.[7] A high pond sits above the

Above the Grasses
AIKEN COUNTY, SOUTH CAROLINA
Known also as high ponds, depression meadows are wetland vegetative communities that are round or irregularly shaped with widely fluctuating water levels. A maple grows in high pond habitat dominated by grasses and sedges.

RIGHT **Kudzu Climbs a Landmark**
AIKEN COUNTY, SOUTH CAROLINA
Near the Janet Harrison Heritage Preserve's southwest end stands an old windmill covered in kudzu. This local landmark has given the preserve a nickname, "Windmill High Pond." Located near Monetta, South Carolina, this thirty-acre preserve protects the high pond vegetative community and several rare plant species occurring within what might be a bay in the making. Far west, as South Carolina goes, bays like this are rare.

water table as true bays do, and though research says no new Carolina bays are developing, some believe this preserve may be a bay in the making. Hard to tell, as no one can fully explain how Carolina bays came to be.

High ponds, distinct depressions on fairly flat, well-drained upland ridges, are among South Carolina's more threatened ecosystems. Only three remain natural enough to merit protection. Some call them "depression meadows," wetland vegetative communities described as rounded or irregularly shaped with widely fluctuating water levels where grasses and sedges dominate.

The Janet Harrison High Pond Heritage Preserve sits off the intersection of Highway 39 and Carolina Bay Trail not far from Monetta. As we drive in at 6:45 A.M., the look and feel of the place resurrect memories of films, books, and *National Geographic* features.

Blackbirds congregate in pines near the preserve, a clamorous welcoming committee. Coils, brambles, and tangles of blackberries fringe the preserve's northeast edge. The bushes are heavy with berries as plentiful as stars in the night. Succulent, shiny blackberries glisten like jewels. Red berries wait their turn for sunshine to do its magic. I eat a dozen big, shiny berries and fill a bag for sprinkling over cereal.

Just beyond the preserve's southwest end stands an old windmill covered in kudzu. The windmill looks like a huge Christmas tree with blades and a vane as the top ornament. Thanks to this landmark, locals refer to the preserve as Windmill High Pond.

Wilderness Bouquet

AIKEN COUNTY, SOUTH CAROLINA

Plants at the Janet Harrison Heritage Preserve rival bouquets and dish gardens. Many rare plants live here: harperella, pink tickseed, Florida false loosestrife, slender arrowhead, dwarf burhead, and others.

The preserve's formal name honors the late Janet Harrison, an ecologist who worked at the Savannah River Ecology Laboratory. Harrison, a native of Greenville, loved South Carolina's wild places, had an inquiring mind, and loved being in the field. A graduate of Presbyterian College, she died in her forties. This wild and beautiful place commemorates her memory and work.

When you go to a Carolina bay, or in this case a "high pond," your senses go on high alert. So much to see. So much to hear. A bobwhite whistles, and though I never hear bobwhites like I did as a boy in rural Georgia, I'm not surprised. In almost every bay I've been to, the bobwhite's sharp, lilting whistle, one of the purest tones you'll hear, sweetens the air. Over the last twenty years, the only quail I've heard have been in Carolina bays. Tells me something about loss of habitat.

The sun rose at 6:14, and at 7:16 it breaks through a low, smoky cloud. At 7:21 it hides again. Rainy weather moved in overnight. Frogs sing. Many wildflowers grow here. We move through the bay. "Man, it's like a garden in here," says Robert.

As many rare plants live here as anywhere in South Carolina: harperella, pink tickseed, Florida false loosestrife, slender arrowhead, dwarf burhead, and others. Amid all the greenery stand snags, proof that something here went wrong for the trees. Standing water drowned them?

"This place is just weeds to a lot of people," says Robert as we head to the interior. He's right. It's easy to ignore or undervalue such a place in our era of manicured lawns and symmetrical flowerbeds. And then there's that old mind-set that swamps aren't good for much.

Thick, shrubby undergrowth makes it difficult to get into the interior. Currents of air carry a heavy, musk-like fragrance, what my mom referred to as "swamp smells."

The birdcalls never let up. Nor do the frogs. It all makes for a pleasant medley. Only an occasional truck mars the peace. Northeast across the bay, the terrain plunges into a water-filled crease, then rises. Sunlight reflects off clouds and lights up the water two hundred yards away, giving it a mirage-like semblance. It brings to mind Hemingway's epigraph to his *True at First Light*: "In Africa a thing is true at first light and a lie by noon and you have no more respect for it than for the lovely, perfect weed-fringed lake you see across the sun-baked salt plain. You have walked across that plain in the morning and you know that no such lake is there. But now it is there, absolutely true, beautiful and believable."[8]

The water is there, for now. No denying it or its frogs' morning music. Several years back, drought allowed pines to invade the edge of the preserve. We've seen drought's effects in a lot of bays. Free of standing water, plant succession takes over, and saplings launch an invasion.

Clouds of gnats hover just beyond my eyes. The wind picks up and blows them away. It blows, too, a familiar refrain my way: a chorale of frogs and birds. The frogs croak and bark in the wet interior where lush sedges grow. The sedges' bright green stands out, and the wind renders them into emerald rivers as they rise and fall. Blue darters fill the interior. Photography and notes: that's our mission, and we accomplish it. The clouds begin to disperse, torn into shreds by winds aloft. We pack up and head back to civilization.

It's easy to focus on the wilderness, beauty, and singularity of a place like this, but what's important is the role it plays as an unusual landform, a conservatory of habitat of rare flora and fauna. To stand here and look across the preserve transcends experiencing an Africa-like place. I'm standing on the edge of now gazing into the past, time traveler that I am.

Not Always Gray

AIKEN COUNTY, SOUTH CAROLINA

Cope's gray tree frogs generally breed in fishless wetlands. Carolina bays that hold water intermittently and don't sustain fish populations suit them. Don't handle them. This frog species secretes a toxin that can bring pain to the eyes, lips, mucus lining of the nose, cuts, and abrasions. Cope's gray tree frogs and the eastern tree frogs appear identical.

Florida Bay and Red Bluff Bay
Francis Marion National Forest, South Carolina
July 24–25

The sun bears down, the humidity's high, and the terrain is rough and remote. To the northwest a cloudbank promises relief that never comes. We're miles from civilization deep in the Francis Marion National Forest. We park beneath high-tension power lines off Halfway Creek Road between McClellanville and Jamestown.

We are on a quest to find colonies of pitcher plants. Most people have never seen these carnivorous plants. To say they're beautiful is an understatement. Let me put it this way: were they easier to access, florists would elevate them to exalted status.

Florida Bay, known also as Wambaw Bay, is a surreal, hauntingly beautiful place. Its colonies of pitcher plants charm the eye. Yellow pitcher plants stand in dense clusters among knee-high red root grass that's thick, green, and luxuriant. The pitcher plants' red throats welcome insects. The plant's unusual blossoms make you do a double take.

Across the savanna wind-twisted pond cypresses conjure up images of acacia trees. A prescribed burn has blackened them. Vegetation is lush. Shorter red-green hooded

The Growth Season
BERKELEY COUNTY, SOUTH CAROLINA
Summer at Florida Bay can be dry. Summer's higher temperatures accelerate evaporation. Plant growth consumes water as well, further drying out the bay.

RIGHT *Florida Bay, Land of Wildflowers*
BERKELEY COUNTY, SOUTH CAROLINA
Florida Bay is among the more beautiful bays when its pitcher plants and wildflowers color the grasses and sedges. The Spanish *florido*—abounding with flowers—makes "Florida Bay" a good name for this bay, known also as Wambaw Bay.

OVERLEAF *Bouquets of Death*
BERKELEY COUNTY, SOUTH CAROLINA
In the Francis Marion Forest, a colony of pitcher plants splashes Red Bluff Bay with autumn colors that rival flowers. Pitcher plants are death made beautiful. Tubular and narrow at the bottom but ever wider to its funnel-like top, the pitcher plant is a doomsday plant for insects and small animals.

A Frame of Young Cypresses
BERKELEY COUNTY, SOUTH CAROLINA
A new generation of pond cypresses grows in the savanna of Florida Bay. Shallow, fluctuating water conditions suit the pond cypress, but the thin soil leads to slow growth.

RIGHT **Carnivorous Colonizer**
BERKELEY AND HORRY COUNTIES, SOUTH CAROLINA
Hooded pitcher plants' flowers resemble upside-down umbrellas rising above the pitchers, an adaptation that protects pollinators from the pitcher's acidic death trap. Colonies of pitcher plants evoke images of alien cities.

Beneficial Bolts

Lighting strikes that burn off woods do good things. They eliminate fuel and underbrush that if allowed to accumulate produce more damaging forest fires. Lightning fires open up the bay's floor to sunlight and kill pests and diseases. Lightning's low-intensity fires damage mature trees less than major fires.

pitcher plants and meadow beauties grow here. Cicadas sing. Yellow milkwort grows here.

It's said this bay is among the most beautiful of the Carolina bays. The vibrant plant zonation—colored bands of grasses and sedges—pleases the eye. The pitcher plants evoke images of an alien city where futuristic skyscrapers rise over a grassy plain.

We make our way into Wambaw Bay, and though we find some pitcher plants, they don't grow in profuse colonies as we hoped. Sedges and wildflowers give us plenty to photograph. Young cypresses stand amid wildflowers.

July 25. The Francis Marion Forest reels beneath angry skies. High winds, heavy rain, and neon-blue bolts strike the land. On Highway 45 we witness a biblical vision. Lightning has struck a pine. Orange and yellow flames send yellow smoke into the gray sky. The pine glows like the sun. Lightning strikes have long produced burns that enhance the quality of life in bays. It took a while, but humankind eventually realized fire can be a good thing as natural communities go.

We turn off Highway 45 onto Chicken Creek Road and negotiate forest service roads that grow increasingly difficult. Finally we can go no farther. We park and slog through

two hundred yards of thigh-high shrubs into Red Bluff Bay, where colonies of plants, *Sarracenia flava,* dazzle us.

From the northeast our friend, the bobwhite, whistles his pure honey call. Every so often a deep crump resounds to the southeast. Thunder.

Friendly Fire
CHARLESTON COUNTY, SOUTH CAROLINA
Here's a fire Smokey Bear would approve of. Prescribed burns in pine flatlands make for healthy habitat by eliminating combustible organic litter, encouraging the growth of understory species that provide wildlife food, better controlling diseases and insect pests, and enhancing native vegetation.

Colonies of pitcher plants stand everywhere. Surely if life exists elsewhere in the universe, this is what an alien plant might look like. The subtle shift of color among the grass zones is a thing of beauty. A bay tree stands in grass, an ecotone green with the color of life. All this beauty is here because people can't convert it to their desires. Imagine a strip mall here. On second thought, don't.

Pitcher plant beauty and seduction reign here. Scents of nectar drift across the currents. Colors and patterns fashion visual lures. Flies, crawling insects, and tiny animals can't resist the pull. By land and air they come to the pitcher plant's lip, which, when moist, proves slippery.

Colonies of Deception

BERKELEY COUNTY, SOUTH CAROLINA

At Red Bluff Bay, pitcher plants, *Sarracenia* family, sport the colors of fall. It's no accident that pitcher plants resemble flowers. That ruse, along with an alluring fragrance, lures insects to their death so the plant gets much-needed nutrients—ecology in action.

ABOVE RIGHT **Where Neighborhoods Meet**

BERKELEY COUNTY, SOUTH CAROLINA

Biologists talk about natural communities using words like ecology, ecosystem, and *ecotone* Here we see an ecotone, a transitional area of vegetation between two different plant communities, a forest and grassland. A characteristic of an ecotone is an abrupt change in habitat, visible here.

I take a long, close look at a colony and can't believe all the flies landing on the plants; they're eager to taste the nectar. Down a throat goes a fly. Once it climbs down the plant's long, funnel-like throat, it's done. After many attempts to escape, exhaustion sets in, and the fly sinks into the acidic water and dissolves, becoming the very thing it craved.

Wildflowers, strange plants, and majestic sedges surround me. Yellow pitcher plants, red root grass, and yellow milkwort splash paint, you could say, here and there across colorful grass zones. There's music to be heard when the frogs sing and cicadas rejoin from the treetops.

Dawn or dusk there's beauty aplenty. At daylight silver spiderwebs bead up with dew, and spice-like scents come off the bay. At sunset western light sets cypresses' crowns afire, and the birds set a frenetic pace as nightfall drops. A bay day has passed.

Lake Waccamaw
Columbus County, North Carolina
October 24

North Carolina's Lake Waccamaw is unique in good and bad ways. It is one of the world's most singular bodies of water. Seen from airplanes, it's been described as a brackish egg dropped into a field of green. It's the largest bay in the Carolinas, but the casual observer will see just another lake fringed by waterfront homes along its north to western shores. When you look at Lake Waccamaw on a map, however, you cannot mistake its elliptical shape for anything but a Carolina bay. When you see the lake in person, it will disappoint you. It looks like a manmade reservoir where nice homes crowd its edge. I would hope the homeowners here understand just how unique this lake is. The state park's observation boardwalk sits along the southeast end. It checks development.

Glittering Cast Nets
BLADEN COUNTY, NORTH CAROLINA
As daylight breaks, grassy zones and pitcher plants glisten with silver films. Spiral webs, funnels, and mesh and sheet webs, silver and sparkly, can be seen everywhere, verifying the large numbers of spiders in bays. Like cast nets, the webs reel in prey.

Driving in, we pass sand rims. It's a beautiful October morning, a pleasant sixty-eight degrees, with ranks of white clouds marching in from the west. Their feathery reflections ripple across Lake Waccamaw. A chorus of frogs sings, and the croak of a great blue heron sounds. I walk onto the long, broad boardwalk that juts into the lake. Two aluminum ladders permit swimmers an easy climb onto the boardwalk. (Alligators swim here, too.) Reeds fringe the eastern shore, where an observation deck juts into shallow water.

Man has colonized this bay. A droning noise comes from across the northern shore, the whine of a leaf blower. At no other bay have we heard a sound from suburbia.

The view across this bay's watery expanse—5.2 by 3.5 miles—makes it seem more like a manmade lake. The tree line edging the bay's distant shore is dark, as if a child drew it in using a thick, stubby pencil. Reeds and cypresses shoot up from the sandy bottom

A Most Unusual Bay

Lake Waccamaw is a big bay, big enough to form the Waccamaw River's headwaters and big enough to attract developers and their passion to build homes along the shores of lakes. But Lake Waccamaw is a bay, in the words of the NC State Park system, that's "one of the most unique bodies of water in the world and one of the greatest geological mysteries."

LEFT **Wave Action**

COLUMBUS COUNTY, NORTH CAROLINA

Waves form patterns in tannin-colored sediments that resemble tiger stripes and footsteps at Lake Waccamaw. Wind and wave action get widespread acceptance as a theory for the origin of bays and their elliptical shape.

red from tannins. The ripple marks in the sandy bottom look just like the stripes on a tiger or a zebra shot over with the red sunlight of a dying day. Ripple marks reveal the influence of wind upon the bay, bringing to mind a theory about the bay's origins.

Lake Waccamaw State Park's website hints of what you can find here: "The cool, tea-colored waters at first appear similar to other lakes in the area, but Lake Waccamaw is one of the most unique bodies of water in the world. You will find here species of animals found nowhere else on the planet, rare plants and endangered animals." One of the rare plants referred to is the Venus flytrap. As we left a woman in the parking lot told us she had been looking for Venus flytraps. She had no luck.

Singletary Lake
Near Elizabethtown, North Carolina, off Highway 53
October 25

Singletary Lake lies within the 35,975-acre Bladen Lakes State Forest. This bay, almost 4,000 feet long, features a pier and nature trail that bring to mind Woods Bay State Park and Heritage Preserve. When mists creeps into the forest surrounding the bay, shafts of light among the trees take on a primeval quality. Some bays change shape, and some shrink. Nearby Horseshoe Lake, a Carolina bay, lives up to its name. Singletary Lake is shrinking as some bays do when plants

Winds across the Bay
BLADEN COUNTY, NORTH CAROLINA
Encroaching tree and plants can fill in a bay's waters. Winds whip white streaks across Horseshoe Lake, an aptly named bay. The burnt orange of cypress needles runs counterpoint to the brilliant green the needles display in spring. Each season brings its own beauty to a Carolina bay.

take hold around their edges. The trees and shrubs establish themselves and act as shock absorbers that cushion wave and current action, letting sediments accumulate, a place for new plant growth. Trees invade the lake, and over time the lake shrinks. Singletary Lake is less than fifty percent of its original size. Its future may be that of a bog.

A mountain laurel relative, the rare white wicky, grows here, as do carnivorous plants. The endangered red-cockaded woodpecker lives here also. Being a state park, Singletary Lake is protected from development and highly accessible. It's a beautiful haven for wildlife and people as well.

Jones Lake
Elizabethtown, North Carolina
October 25–26

Dusk. We set up camp in Bladen County at Jones Lake State Park, the centerpiece of which is a Carolina bay, Jones Lake. Spectacular sunset. The cold night is clear, and no city lights pollute the night sky. A million stars salt the firmament, and come dawn a royal spectacle arrives. Dew coats everything with silver lacquer as morning breaks pink and rosy in the high thirties. Gray fog rises from warm waters with a pink sunrise as a backdrop. From the bay's edges, wisps of fog gravitate toward the center.

Beyond the fog over the tree line, soft layers of gray, blue, and pink reflect colors of the sun, which hides beneath the horizon. From the dense fog, an explosion sounds,

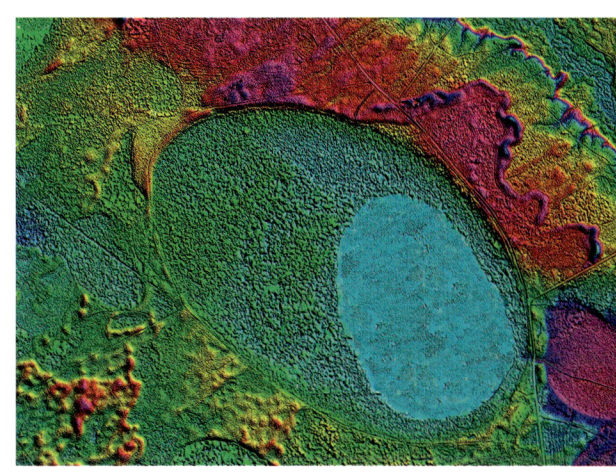

erupting feathers. Wings whip the water, and Canada geese launch down their watery runway and take flight to the west. The fog thins, and the sun breaks through, letting the colors of the day assert themselves. The sun climbs and burns the fog away. The magical morning ends.

The water's still as glass, resurrecting a song that I think is about a Carolina bay in North Carolina. In Fleetwood Mac's "Hypnotized," songwriter Bob Welch references an odd pond. Said Welch, "The 'strange, strange pond' with 'sides like glass' refers to a strange anomalous depression in the North Carolina woods near Winston-Salem, North Carolina, which a friend told me about, which, at the time, freaked him (and his dirt-biking

Fog Rising

BLADEN COUNTY, NORTH CAROLINA

Set amid a rim of green forest, Jones Lake shimmers pink. No stream, creek, or spring feeds this bay, just rainfall. The night before, chilled air heavy with dew coated the land with silver lacquer. A million stars salted the night, and come dawn a royal spectacle arrived. Aurora, goddess of dawn, spreads her gown of gray over coral waters.

TOP *Atypical Cypress*

BLADEN COUNTY, NORTH CAROLINA

This strange, strange cypress stands in Jones Lake. With a shoreline of 2.2 miles, it's a sizeable, water-filled bay with stands of cypress that tower over the water. You will see what appears to be stunted cypress as well.

Foggy Liftoff

A flight of Canada geese quietly clears the fog over Jones Lake. *Branta canadensis,* a large goose, is more at home on a bay than a city park. For whatever reason no honking took place in the bay.

Bay Beauty

BLADEN COUNTY, NORTH CAROLINA

As in an impressionistic painting, shifting light and shimmering water render buttressed cypress into art. Throughout the seasons a Carolina bay provides photographers a gallery of subject matter.

buddies) out."[9] The song describes a pond "in the thick of a forest without a road / And if any man's hand ever made that land / Then I think it would've showed." Welch's lyrics would suit meteorite theorists.

Jones Lake bay lies upon the land like a lost pendant. Set amid a rim of green bay forest, it's tranquil and shimmering, a jewel of a natural area. No stream, creek, or springs feed the lake. It relies on rainfall, and like its sister bays, its long axis runs northwest to the southeast. It's a classic Carolina bay, of which North Carolina state park literature

Autumn Descends

BLADEN COUNTY, NORTH CAROLINA

Autumn falls upon Jones Lake. While fall's palette brings vibrancy to cypress, hardwoods, and pitcher plants in the bay, it also brings a time when leopard frogs and American bullfrogs prepare to hibernate, with occasional warm days bringing them out.

says, "Venture to Jones Lake State Park and view one of the greatest geological mysteries of the eastern United States—the phenomenon of the Carolina bays. Adjacent to the Bladen Lakes State Forest and home of two natural lakes, Jones and Salters lakes, the 2,208-acre park is a nature lover's delight."

Not even fog can mute fall colors at Jones Lake, a bay not far from the Cape Fear River. This watery region had to be a haven for the Great Spirit's children and creatures.

Cathedral Bay
Olar, South Carolina
November 10

This day also begins in fog, which thickens as we drive down Highway 321 and approach Norway, Denmark, and Sweden, a Scandinavian series of towns here in the Southeast. A profusion of bays exists here in faux Scandinavia. If you look at a map, you'll see that a bounty of bays pockmarks the land on either side of Highway 321, known as the Carolina Highway and Savannah Highway.

We turn onto Highway 64 and soon park at the edge of Cathedral Bay, not quite two miles from Olar. We're here because this natural heritage preserve provides a classic example of the Carolina bay phenomenon. It's cool, fifty degrees, and overcast. Aerial photographs reveal that fifty-eight-acre Cathedral Bay, once known as Chitty Pond, appears to be slightly offset from a "ghost bay," an antecedent or previous bay.

Cathedral Bay is known for its pond cypress waters. Heavy rains have the trees standing in deep, black, motionless water. Owing to a lengthy dry period, sedges have given way to thick tangles of holly and

Tangled Waters
Drier times permit sweet gums and hollies to invade Cathedral Bay. Let the waters stand long enough, however, and the invasives will die out.

Red Berries for Bluebirds
BAMBERG COUNTY, SOUTH CAROLINA
In winter birds, squirrels, and deer will turn to
holly berries for sustenance once preferable
berries have been eaten. Bluebirds especially like
holly berries.

RIGHT ***Winged Fruit***
BAMBERG COUNTY, SOUTH CAROLINA
Think of a maple's winged seeds, samaras, as a
bit like helicopters. Come spring, their spiral-
ing flight drifts them to a spot, ideally, where
they can germinate. The samara is all about
seed dispersal, and in bays where drought
has set in, they germinate and readily grow
where sedges once grew.

maples, which stand among ranks of pond cypresses. Dry times, however, produce carpets of magnificent grasses that lie like fleece at the trunks of cypresses. Myrtle-leaf holly is common in the shrub layer, with wax myrtle and buttonbush less common.

We wade into the interior. The water is a cool sixty degrees. Fallen trees and submerged logs make the going tough. Lots of bird life but few insects and no snakes. Now and then things leap into the water. They have an uncanny way of jumping when we aren't looking.

There was a time when grasses and sedges softened the bay's interior and surrounded the base of buttressed cypresses like batting. Not now. Holly and maples grow where sedges once grew, but I hope the heavy rains drown out these invasives.

The South Carolina Department of Natural Resources manages Cathedral Bay through its Heritage Trust Program. You can find this accessible, extraordinary pond cypress bay near the junction of SC Highway 64 and US Highway 301 as well as just off 321 via 64.

Fall is in command of the bay. Cypresses are a subtle burnt orange. Brown grasses and sedges resurrect images of my grandmother's brooms she fashioned from broomstraw. Brown toothache grass holds a secret: it numbs the mouth. Here and there zones of green grasses flourish, and just back of the brown zones stand dry pitcher plants.

Lime green pitcher plants jut over brown sedges. Dead pitcher plants stand side by side with live ones. Deer trails through the savanna flatten dry grasses. Down close to the ground, you can find small yellow wildflowers shooting up among turkey oaks. Dry dead leaves and grasses surround white

Nature's Painkiller
BERKELEY COUNTY, SOUTH CAROLINA
Toothache grass (*Ctenium aromaticum*), a warm-season, perennial bunch grass, grows in some bays. It numbs the mouth and creates a tingling sensation, thus giving Indians a remedy for toothaches.

LEFT **Herbivore Highway**
BERKELEY COUNTY, SOUTH CAROLINA
Deer trails run through Carolina bays, where deer find cover, food, and water—their basic needs.

RIGHT **The Colors of the Earth**
BERKELEY COUNTY, SOUTH CAROLINA
This piney savanna may not seem diverse, but it is. A sunlight-rich, open sky, grass, shrubs, wildflowers, and wildlife species fashion a vibrant ecosystem.

daisy-like flowers. Subtle colors, as if a watercolorist used the palest hues of lavender, gold, wine, purple, browns, and greens. Cocoa and burnt sienna. The bay appears washed out compared to June and July.

Charred pieces of wood and smudged bark confirm a prescribed burn passed through. Fire partially melted a Forest Service location sign, N 6, tacked onto a pine. A distinct edge defines where the pitcher plants end and brown sedges take up. Here two types of habitat meet. Pitcher plants have given up the ghost. Their vitality sapped, all are brown save two that retain vivid lime hues. When I touch the brown, dried-out pitcher plants, they rattle like paper. The ground is not dry, though. Recent rains produced seeps; the oozing wetland makes sloshing sounds as I walk.

Hardwoods don't grow well here, but ferns do. Lots of wildflowers, free of a canopy's shade, flourish. Dainty, sublime wildflowers abound. Sulphur butterflies light on small yellow blooms. Maize on mustard. In the center of the savanna, brown grasses assume elliptical form, an oval within an oval.

As I walk through the grasses, they swish, and every ten feet giant grasshoppers take flight with a rattling of wings. Lots of white top sedge. From the northeast a woodpecker sounds as if he's knocking drumsticks together. To the south starlings raise a ruckus. Three manmade noises mar the symphony: deer hunters' gunshots, an airplane, and trucks out on Halfway Creek Road.

I cut a pitcher plant apart, curious to see its business end where insects meet their maker. A musty dead fish odor rises from what looks like fine silt. The pitcher plant has manufactured soil on its own.

Despite its remoteness, people have come close to disturbing Florida Bay. Skirting the perimeter of the bay is the Wambaw

Fire's Passage
BERKELEY COUNTY, SOUTH CAROLINA
Lightning-sparked wildfires long burned off bays. Today forestry personnel set bays afire to kill back shrubs and hardwoods so native grasses and wildlife can thrive. Blackened pine bark is a sign that luxurious growth is coming, diverse growth that's a food source for wildlife species.

Primeval Plants
BERKELEY COUNTY, SOUTH CAROLINA
Were you to inspect the soil beneath these ferns, you'd see the charred remains of a controlled burn. Ferns are among our older plants, with a fossil record dating back some 390 million years.

Cycle Trail, a gas pipeline, a high-tension power line, and a highway. A fence skirting the bay just beyond the pipeline keeps out ATV riders who wheeled through the bay in the past.

What's it like to view Florida Bay from the power line right of way in autumn? You see a distant line of rust-smudged trees, cypress needles in fall clothing. The treetops look as if an artist took burnt-sierra chalk and rubbed a thick line against a gunmetal sky. Here and there you see subdued green, a fading summer memory. The land plummets from the power line right of way and falls out of sight—as a legendary depression should.

Needles in Transition
Fall touches the bay, and longleaf needles respond to the tilt of Earth's axis.

RIGHT **Over the Boardwalk**
LOWNDES COUNTY, GEORGIA
Near Valdosta, Georgia, a 2,600-foot wooden walkway takes visitors to a rebuilt fire tower 54 feet above a watery prairie that's part of an 18,000-acre wetlands complex of bays and swamp second in size to the Okefenokee Swamp. Adjacent to Moody Air Force Base, the Grand Bay Wildlife Management Area encompasses 5,874 acres of land under license from the U.S. Air Force.

Grand Bay
Lowndes County, Georgia
November 18

A drive of seven hours through the backroads of Georgia—that's how long it took to get to Grand Bay near the Florida state line. Valdosta, Georgia, to be precise. Grand Bay edges Moody Air Force Base. Grand Bay, 1,350 acres, is part of an 18,000-acre wetlands complex of Carolina bays and forested swamp. It's second in size only to Okefenokee Swamp and lies in a geographic province known as the Flatwoods.

Driving in, we spied a wake of buzzards dining on a deer carcass. Buzzards were in the trees and all over the ground—the most I've seen in one spot. We made our way to the Grand Bay Wetland Education Center and parked. We walked the boardwalk (2,600 feet) and halfway down it came across a six-foot alligator. It lay there. No movement. Its tail lay among water lilies; its head faced a patch of fall-struck vines.

At the end of the boardwalk, the observation tower (a restored fire tower) rises over the bay. Some sixty-six flights later, we stood fifty-four feet above Grand Bay, overlooking

Drying Out

The anhinga, known also as snakebird and water turkey, swims underwater in search of fish. After a hunt it dries its wings, as its feathers have less oil than other birds.

Armor Plate

LOWNDES COUNTY, GEORGIA

North America's largest reptile, the American alligator, possesses a tough hide, one prized for belts, shoes, purses, and jackets. Once hunted to dangerously low levels, gators made a comeback thanks to protection, and hunting today is highly regulated. Gators have no vocal cords but roar by sucking air into their lungs and blowing it out.

open prairie and cypresses. At 5 P.M. sharp, trumpet music from Moody Air Force Base, four miles distant, sounds retreat for the day, the notes clear and surreal. "The Star-Spangled Banner" plays, dreamlike, wavering. On cue a broad-shouldered hawk calls out and, like a U.S. Air Force Raptor-22, skims the tops of cypresses below. It's different to see a hawk below soaring over burnt-orange cypresses casting images across dark blue water flush with flotillas of lily pads.

The boardwalk cuts through a cross section of communities—a mosaic of wet savannas, shrub bogs, cypress-gum ponds, and prairie and black gum cypress swamps similar to Georgia's Okefenokee. The abundant diversity of wildlife is similar to Okefenokee's as well. A small percentage of the area is in mixed live oak and pine. Gopher tortoises and indigo snakes live here.

Grand Bay isn't quiet. A chorus of pig frogs and cricket frogs serenades the bay. Gators boom out. Had we been there in spring, prothonotary warblers might have added their songs to the mix, flashing bursts of gold across blue dark waters.

Wading birds: cattle and great egrets; great blue, little blue, green, and tricolor herons; and white ibis. An anhinga takes a break from feeding in the marsh and dries out its wings. Migratory greater sandhill cranes usually arrive at the marsh in mid-November and remain until mid-January. We saw none. Too early, perhaps. You can see white-tailed deer, raccoons, bobcats, gray foxes, opossums, and armadillo here.

Locals say the bay is spectacular beneath a full moon, that it's a beautiful time to stroll the boardwalk to the observation tower. Imagine a moonlit night looking down on the nightlife of a bay. Imagine hearing the calls of owls and whip-poor-wills floating up fifty-four feet into the night.

Banks Lake
Lanier County, Georgia
November 19

Early morning. We set out for Banks Lake, a bay near Lakeland, Georgia, a beautiful, serene setting. Though colonized by people in the usual ways, it retains its wilderness character, this cypress-filled bay with mists turned pink by the rising sun. In short order the mists dissipate, and pink waters turn blue. Here and there cypress knees lean to and fro like meerkats. Those, along with heavily moss-draped cypresses and the wide expanse of blue water with dawn-burnished cypresses on the far edge, ironically bring the Kalahari Desert to mind.

Banks Lake, one of four Carolina bays in the 13,000-acre Grand Bay–Banks Lake complex, neighbors Banks Lake National Wildlife Refuge and is one of the refuge's more notable features. This shallow 106-acre blackwater lake supports fish species and other aquatic animals. More than 150 years ago, man's water-control devices changed the Grand Bay–Banks Lake complex, altering natural drying and fire cycles, which affected naturally occurring plant and animal communities.

Some of Banks Lake is open water; much of the interior and boundary is marshy and dense. A highway skirts its edge, and a boat

OVERLEAF **Dawn's Glow**
LANIER COUNTY, GEORGIA
Banks Lake is known for its moonrises and sunsets. Witnessing sunrise at a Carolina bay is an experience too few enjoy. Mirror-like water, lily pads still as stones, dawn's early light, and pink cypress knees that preen like meerkats provide ample sights.

Ranks of Russet Cypress
LANIER COUNTY, GEORGIA
At Banks Lake draperies of Spanish moss bespeak of fall, a time when visiting a bay is a bit more comfortable—fewer insects and tolerable temperatures. For many bays it's a time when water levels are at their lowest.

RIGHT **Like Green Stepping Stones**
LANIER COUNTY, GEORGIA
Lily pads create a stepping-stone-like walkway to cypress at Banks Lake. Most of the plant hides beneath the water. Lily pads provide shelter for fish and cool the water via shade. Frogs perch on them, out of sight of underwater predators. Dragonflies use them as helipads. Lilies also produce oxygen that lets fish breathe and beneficial bacteria thrive. The pads attach to the soil via a long stem. The roots are rhizomes, tubers filled with nutrients. Water lilies live only in shallow, still fresh water.

ramp slips beneath its waters. At dawn a fisherman puts out and etches a feathery wake across the mirror-like water. Morning traffic picks up, and dawn's magic spell surrenders to the workday. Even so, cypresses' tops flame with eastern light as their buttressed trunks hide in shadows. Later lily pads as big as plates will serve as helipads for dragonflies. A tranquil feeling prevails despite the growing traffic.

Carolina bays dot the landscape from New Jersey to Florida. They can be the size of a backyard or many acres. Regardless of size, they're among Earth's more mysterious landforms. Most occur in Georgia and the Carolinas. If you've never been to a bay, you now have some idea what they are like, though all are different.

Once drained at will, bays now enjoy more understanding. They continue to do what they've always done: control floods, filter water, shelter wildlife, and give people a place to sort out things.

Escape civilization. Visit a bay and sort things out. See the rich diversity of wildlife. Take a camera and record your own dispatches from the field.

We hope our book brings more attention to their beauty, their usefulness, and the need to reclaim the ones we can while preserving the few undisturbed bays that remain. Once you've seen a few bays, we believe you will agree.

A Square Foot of Grass

In "East of Eden," Dickey quoted Philip Wylie, who often wrote about ecology: "The sum total of human works, the artifacts of savages, barbarians, medieval men and modern, all cities and towns, every hut, hovel, skyscraper and temple, all steel bridges, everything man has made to use since the first stone tool or wooden club—does not equal, in all parts put together, the achievements of the life forms of plant and insect in a square foot of grass."

The complicated mechanical movement of the finest Swiss watch pales in comparison to a square foot of grass. Soil type, sunlight, the presence or absence of water, available nutrients, flora, fauna, and how everything works together is complex, near mysterious. Everything aboveground, in the ground, below the ground, in waters, in the sky, and beneath the sky must mesh. A Carolina bay and its inhabitants depend on everything coordinating—life rises from death and from life death must follow, and the great cycles go on creating a magical, mystifying world.

Rainfall Dependent

Carolina bays depend on rain for their water. Thus they may be inundated, partially filled, or dry. Many bays retain rainwater—some a great deal of the time, some now and then, and some year-round. Carolina bays represent the coastal plain's dominant freshwater wetland feature. Think of Carolina bays as basins open to rain set into the coastal plain.

Though a few bays have streams running into them, most bays do not receive water from streams and springs. It falls from above and pools in bays' depressions. Water seldom flows from them (unless ditched), but it evaporates and escapes into the atmosphere through transpiration from the leaves of plants. Living things consume water

Longleaf's Long Journey
CHARLESTON COUNTY, SOUTH CAROLINA

A longleaf begins life in a bay's sand rim. The slow-growing longleaf pine depends on fire to reproduce. Historically lightning-strike fires cleared away leaf litter so seeds could germinate. The longleaf pine goes through a grass stage, a large clump of needles, during which it develops its taproot. These slow-growing trees live for more than three hundred years.

Black Waters

A swirl of autumn leaves and patches of blue sky brighten Woods Bay's dark water. Buttressed trunks reveal a dropping waterline.

through drinking and the eating of plants and prey. The amount and duration of water in bays prove valuable to many species in different ways.

Bays' unique combination of natural features supports a wide range of habitats. It's not unusual for a bay to feature cypress pond, hip-to-chest-deep blackwater, boggy, mucky areas, pocosins, savannas, and desert-like sand rims. From pocosins to sand rims to wet areas and savannas, bays provide much-needed habitat for common and rare species and a refuge for species suffering habitat loss and diminishing numbers such as longleaf pines.

Bays provide habitat for flora and fauna that possess relationships, and seeing relationships in action drives home the point that vastly different species interact in and around bays as they pursue the business of living. Pitcher plants come to mind at once.

Sarracenia

With their vivid colors and alluring fragrance, pitcher plants, members of the sarracenia family, illustrate ecology in action. Pitcher plants resemble flowers by design, and they are among the more beautiful plants in some bays. On top of that, the pitcher plant may well be the world's most mysterious leaf. Tubular and narrow at the bottom but ever wider to its funnel-like top, the pitcher plant is a deathtrap. Because they typically grow in soggy, acidic soil that's weak in nutrients, they commandeer nutrients from creatures that fly and crawl about. That's where their modified leaves come into play. Pitcher plants lure insects and small animals with fragrant "nectar." Looking like flowers enhances the deception.

Pitcher plants are otherworldly, inventive plants. Not even rain can weaken their deadly effectiveness. The lid or hood that covers some pitcher plant species deflects rain, ensuring that rainwater doesn't dilute their deadly pools of acid. Other pitcher plant species welcome the rain as a drowning pool for insects.

For small creatures a pitcher plant colony constitutes a death zone. A sweet fragrance emanates from the world's most mysterious leaves, drawing insects inside the tubular leaf, where a natural wax proves slippery. Downward spines make it easy to descend, but once inside, insects can't climb out. The once-helpful spines now present formidable barriers. Exhausted by escape efforts, insects and small animals fall into liquid at the bottom of the leaf. Acids go to work. The insects slowly dissolve, becoming the very juice they hoped to drink. A diet of insects provides the pitcher plants what impoverished soils cannot—nitrogen for amino acids and sustenance.

Field of Trumpets
Colonies of pitcher plants stand ready to help control insect populations in the Francis Marion National Forest.

Ambush!

A nursery web spider stares down the throat of a yellow trumpet pitcher plant. Spiders spin webs across pitcher plants' throats, taking advantage of the plant's appeal to insects and small creatures. To avoid falling into the throat, spiders attach a silk thread to themselves and the plant's exterior, a lifeline of sorts.

RIGHT *A Sticky Situation*

Remember flypaper? Sundews are nature's version. Yet again a sweet fragrance lures insects to tentacles with dew-like glue drops at their tips. Once the insect is hopelessly stuck, tentacles wrap around it and kill it. Digestion then begins.

This life-and-death drama is not lost on spiders. When the sun breaks over Red Bluff bay, spiderwebs glisten among the sedges and grasses. Spiral webs, funnels, and mesh and sheet webs, silver and sparkly, glitter everywhere. Along the edges of grasses stand colonies of pitcher plants, their crimson and lemony colors contrasting against the green grasses. Look closely, and you'll detect a film of silver covering pitcher plants' throats. Opportunistic spiders build webs across the opening to capture insects answering the siren call of fragrant nectar. The spiders, competing for limited resources, cut insects off at the pass. Green lynx spiders stalk the tops of pitcher plants as well, seizing curious insects with lightning speed.

Wolf spiders build their nests near sundews, a carnivorous plant whose sticky secretions trap insects attracted to its sweet-smelling but glue-like nectar. At first contact the insect is doomed, and soon the plant's enzymes will digest it.

Symbiosis in action: pitcher plants, spiders, and the pitcher plant midge share a beneficial relationship. While some believe spiders harm the pitcher plant by building webs across its throat and competing for prey, a theory holds that spiders' nitrogen-rich excretions benefit the plant. As for the pitcher plant midge, it lives in the purple pitcher plant. It helps break down insects, and when it excretes nutrient-rich feces, it does the pitcher plant a favor. In ecology what goes around comes around. Snails and slugs prey upon pitcher plants, and snakes and shrews prey on snails—the proverbial dog-eat-dog world.

Here's an interesting symbiotic twist. While carnivorous plants depend upon the amino acids from dissolved insects for survival, some species depend solely upon carnivorous plants for reproduction. Take the yellow trumpet pitcher plant, *Sarracenia flava,* for example. Some mosquito and other inspect species gladly enter the plant and avoid being trapped. They find the plant's rainwater pool a good place to lay their eggs.

Inside the Legendary Venus Flytrap

The tremendous range of habitat at Lewis Ocean Bay Heritage Preserve translates into a great diversity of life: Venus flytraps, yellow pitcher plants, native orchids, black bears, bald eagles, and red-cockaded woodpeckers. Twenty bays and dozen or so habitats exist in this 9,393-acre preserve. Differing depths of water, peat mats, and sand rims support a diversity of life that includes water lilies, sedges, red bay, sweet bay, butterworts, sundews, gallberry, white- and yellow-fringed and rosebud orchids, and the rare Wells pixie moss.

Venus flytraps, *Dionaea muscipula,* merit special mention. First it bears noting that it is on the endangered species list in North and South Carolina. It is illegal to take these

Cautious Bloomer

A bloom tops off the long stem. The height keeps insects pollinating the plant from being trapped by it, thus making reproduction more assured. Plants like the Venus flytrap depend on fires to keep its habitat open and sunny.

plants from the wild or to disturb their habitat. The Venus flytrap, a perennial flowering plant, grabs headlines for being carnivorous. Let's go inside the plant to see how it kills.

Its "trap" consists of two hinged leaf lobes. A thick vein running down the leaf's center creates a hinge. The leaf stays open until tiny hairs are contacted, but contact has to be just so. On the lobes' inner surface, hairlike projections, trichomes, trigger the lobes to snap shut but only when prey contacts them twice. Touch a trigger hair once, and nothing happens. Touch one trigger hair twice or touch two different trigger hairs, and a doomsday scenario begins. In twenty seconds the trap snaps shut. This mechanism keeps the trap from wasting energy on rain and other nonnutritious things.

Touching a hair fires an electrical signal along the trap's surface. Touching a hair the second time sends a confirmation current causing cells in each lobe's center to lose water, eliminating the pressure that keeps the lobes rigid and separate. The lobes snap together. After the trap snaps shut, the leaves turn into a stomach. Panicking insects make frenzied attempts to escape, but their wriggling closes the trap tighter.

The plant, like most organisms in nature, doesn't waste energy on debris falling into its paired leaves. The trap only shuts when an insect touches the trichomes more than once. The hinged traps' spines interlock when the trap shuts. Think of them as bars in a jail cell.

Digestive enzymes soak the doomed insect. The goal? To liquefy the insect so the plant can absorb it. Digestion can take five to twelve days. After two or three meals, a trap blackens and dies, and new traps take its place.

The plant depends on insects for pollination, but what keeps the plant from eating would-be pollinators? An evolutionary adaption. The flytrap's white cluster of flowers sits atop a tall, slender stalk well away from the trap. Long may it reproduce. It's estimated that flytraps can live up to twenty years, perhaps longer, in the wild. That's a lot of dead flies.

Habitat Rich

Bays' life-and-death scenarios, diversity, and ecological relationships come down to the habitats they support. Depending on its situation, a Carolina bay may support plants that include herbs, shrubs, trees such as their namesake bay trees, and grasses that include rushes or sedges, and then there are ferns and vines.

Average water depth and soil type exercise significant influence on the types of plants living in and near Carolina bays. Expect to see black gum, sweet gum, magnolia, bald cypress and maple, and shrubs such as sumac, buttonbush, gallberry and red bay. Wet Carolina bays commonly feature water lilies.

Many bays feature sand rims along their southeastern edge, thus providing desert-like conditions for species needing dry habitat. Into this dry habitat come many wildlife species. You'll spot their tracks among the hardy oaks that live there.

Fluctuating water levels provide a wide range of habitats from briefly flooded lands to areas that stand in water for long periods. Autumn brings the lowest water level; spring the highest. Some Carolina bays stay wet all year; others fill, then dry as seasons bring rain or dry spells. Rainfall is the key. At

Animal Tracks
Deer and other wildlife species pass through sand rims. Like a fresh layer of snow, a bay's white sands reveal the comings and goings of wildlife.

the Savannah River Site during a thirteen-year period, Rainbow Bay held water 5 days one year but 280 days another.

Green Highway
Once again cypress reveal how water levels fluctuate here where a green veneer of water lilies paves the water for frogs, dragonflies, cousin damselflies, and other small creatures that find the pads inviting.

Lack of rain combined with summer's higher temperatures and water-consuming plant growth speed up evaporation, and affected Carolina bays dry out faster. When spring brings rain, the wettest bays feature open water and aquatic plants such as water lilies, grasses, and rushes. Some bays have an open canopy of pond-cypress trees beneath which grow maiden cane, sedges, and other herbaceous plants. In spring you'll see bladderwort's yellow blossoms.

A Two-Way Street

Water's duration is a double-edged ecological factor. Carolina bays provide crucial habitat to species needing a watery environment for part of their life cycle. Bays that dry out now and then make it tough for fish to live in them. That, however, has an upside. It's easier for salamanders to breed and survive where no fish exist to eat them. Dry periods assure better protection for the aquatic stages of frogs, toads, and salamanders. Predators love amphibians' eggs and easily pick off slow-moving tadpoles and salamander larvae.

Carolina bays continually filled with water would develop fish populations, bringing more predation to bear on amphibians. Species suffering habitat loss such as the rare flatwoods salamander find refuge in some Carolina bays.

Steve Bennett is most familiar with the ecological needs of amphibians. In the Sierra Club of South Carolina's *Congaree Chronicle*, he wrote about his time working with Carolina bays.

> I spent over a decade studying and cataloguing Carolina Bays in South Carolina, along with my close friend Dr. John Nelson. As long as you don't want to know what caused or created Carolina Bays they are relatively easy to understand. They are geomorphic features that share several characteristics in common: they are elliptical or oval basins and, for the most part, the long axis of the ellipse (or oval) aligns along a northwest-southeast direction. The other really important characteristic they share, or most of them share, is that the basins fill and dry with the rainfall cycles. This characteristic throws them back into the eco-semantic torrent many of us had to navigate as we went about our studies and research. Most Carolina Bays are functionally similar to an entire suite of wetlands that go by, or have gone by such names as "high ponds," "wet

CLOCKWISE FROM TOP

The Eyes of a Predator

AIKEN COUNTY, SOUTH CAROLINA

Southern leopard frogs patiently wait for prey to come within striking distance, then pounce on it. They eat beetles, ants, flies, worms, smaller frogs, birds even. Their croaking voice sounds like a full balloon being rubbed.

Forest Habitat

BERKELEY COUNTY, SOUTH CAROLINA

Deep woods surround the bays in the Francis Marion National Forest. The land running up to a bay is vital in that many species live there, and these species go into and out of the bay as natural cycles dictate.

Gelatinous Clutch

AIKEN COUNTY, SOUTH CAROLINA

A mass of frog eggs' shape differs depending on the species. Leopard frogs' egg masses look like softballs. Chorus frogs lay eggs in a long mass. Toads lay eggs in long strings in shallow water. Each black dot consists of cells that will become a tadpole. Fish and other predators feast on frog eggs.

weather lakes," "flatwoods ponds," "bays," "vernal pools" and other somewhat colloquial designations. From an ecological perspective, we have tried
to fit them all under the big and somewhat porous term "temporary, isolated
freshwater wetlands." For the most part, this name works okay. Sure, some are
relatively permanently flooded or maintain water for long times. And some have
surface connections via small streams that flow when water in the basin is high
and may flow into other such wetlands or when nearby streams and lakes over-
flow and excess water follows contours in the land to these "isolated" wetlands.

Yes, there are exceptions to almost every situation governing these temporary waters. More often than not, most of them spend some time completely dry (possibly a few years) and some amount of time very full (once again, possibly a few years)—that, as they say is the heart of the matter.

Steve Bennett, again, writing in the Sierra Club of South Carolina's Spring 2017 *Congaree Chronicle*, addresses the dual aquatic personality of these wetlands. "Isn't it better for a wetland to be wet all the time? The answer is: not necessarily, especially if you are an amphibian like our gopher frogs. This rare frog, along with several other rare amphibian species and a host of less rare, some even common species select exclusively, in the case of the gopher frog, and preferentially, in the case of other species, these temporary ponds as breeding sites."

The explanation for why frogs prefer the bays is relatively simple: to avoid critters that might eat their eggs and, in particular, their larvae. Adds Bennett, "A number of fish species can find their way into a temporary pond when the adjacent streams and permanent

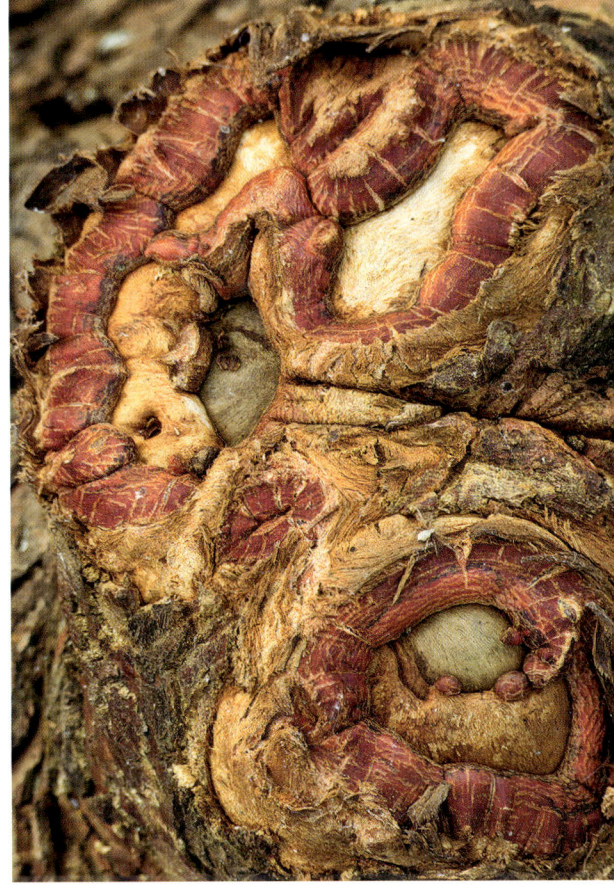

ponds are flooded and take up residence in the pond for a period of time. I have seen very young red fin pickerel, for instance, swimming like crazy along flooded deer trails headed straight for a flooded Carolina Bay. I have sampled temporary ponds that have been flooded for two to three years and found them supporting robust populations of red fin and chain pickerel, mud sunfish, flyers and other sunfish species and other predatory fish species. When the pond dries

Mystery and Beauty

AIKEN COUNTY, SOUTH CAROLINA

In bay-rich Savannah River Site, aptly named Enchantment Bay provides enchanting beauty. Cypress knees, function debatable, serve up a bit of mystery as well. Here we see lily pads, blackwater reflections of blue sky, cypress knees, and red foliage focal points.

RIGHT **Strange Wood**

KERSHAW COUNTY, SOUTH CAROLINA

Cypress knee details look a bit like a grotesque mouth and eye. It remains unknown what cypress knees' functions are.

Monarch's Toxic Friend

BERKELEY COUNTY, SOUTH CAROLINA

A Florida Bay monarch butterfly sits upon orange milkweed, which contains poisonous compounds. Monarchs' unique adaptation lets its caterpillars feed on this plant and thus absorb the toxins, providing protection for caterpillars and adult butterflies against birds and other predators.

Crotalus horridus

AIKEN COUNTY, SOUTH CAROLINA

Despite its frightening name, the canebrake rattler is beautiful. Known also as the timber rattler, this snake spends most of its time coiled in ambush positions. Timber rattlers appear a bit different in South Carolina's mountainous corner. The piedmont-coastal form exists in the region known for Carolina bays. The mountainous form has a yellow to black background, whereas the coastal form has a background color ranging from light tan to orange.

out completely, these fish die and become a nutrient source for the pond. Additionally, just as predatory fish 'disappear' so do the predatory aquatic larvae of insects such as dragonflies and others species."

The elixir of life comes and goes. South Carolina Department of Natural Resources hydrologist Scott Harder noted that "some Carolina bays may only hold water for several weeks of the year; others may hold water year-round except during severe drought, while a few bays can hold water even during severe droughts. These wetlands' ability to retain water along with their varying hydroperiods provides interesting habitat conditions."[1]

"Hydroperiod" refers to the seasonal water-level pattern created by the combination of the "water budget" and the "storage capacity" of the wetland. Water budget refers to the net result of all water flowing into and out of a bay. Winter and spring are wetter times when rains come and evaporation is less. As for bays that experience prolonged dry periods, they may have species that survive in an extended state of torpor or dormancy.

Like other wetlands, bays do good things to water. In addition to supporting biodiversity, they filter out impurities, trap heavy metals, and process microbiological elements. Bays also store storm water runoff. Many bays' high-organic soil acts like a sponge. Thus when the rains come, soil holds the water and releases it slowly. Bays help control erosion, improve water quality, and give people a place for recreation and the observation of nature.

How poor the land would be without bays and their life-giving water. Years ago when I worked on a film that explored bays' origins, I extolled them as wildlife oases. The metaphor of an oasis was not a good choice, then or today. "Use the metaphor of saying a bay is an oasis with care," said Linda Lee, wetlands ecologist. "The land running up to a bay is vital. It's vital in that many species live there, and these species go into and out of the bay as natural cycles dictate. Consider for example the great migrations of salamanders. Marbled salamander adults begin their migrations to breeding sites such as Rainbow Bay as early as August. Migration peaks in October."

Lee advised me to think of bays as a pond on a farm or a fountain in a garden. "Bays provide an option to wild things that's much needed," she said. "Remember that bays change from year to year, and wildlife must have other bays nearby in diverse conditions to carry out their life cycle and to meet their needs. Bays' varying hydrologic cycles, while challenging species at times, also provide seasonal advantages."

She added that it's important to realize that bays are not static. One year they may be good for gopher frogs. The next year, maybe not. Preserving a bay doesn't assure that you save what's in it, and clear-cutting a forest up to the edge of a bay isn't good. It harms amphibians and reptiles that visit the bay during breeding cycles.

Thanks to their varying water levels, bays provide more diversity than other wetlands, and the uniqueness of bays themselves leads to an even richer assemblage of life. The bays also support a long list of plants, including rarities such as bog bluestem, Savannah iris, Boykin lobelia, Canby dropwort, mock bishop-weed, parrot pitcher plant, and many others. Then, too, there's the mysterious cypress knees, functions unknown.

Cypress knees occasionally add to their mysterious nature with unusual appearances that work the imagination, but they are part of the great diversity of plants that give rise to a great diversity of wildlife. The fauna, from small to large, found in Carolina bays includes aquatic invertebrates, reptiles, zooplankton, amphibians, and mammals such as the marsupial opossum. By far most abundant are the frogs and salamanders. Rare animal species associated with Carolina bay habitats include the Eastern indigo snake, gopher tortoise, black-banded sunfish, carpenter frog, and the round-tailed muskrat, among others. At Lewis Ocean Bay, Venus flytraps and black bears have something in common. They are two of the most remarkable carnivores living there.

Carolina bays—beautiful dish gardens—aren't the "worthless swamps" they were once viewed as with disdain. Appreciation for them grows. Noting why researchers have been slow to study bays, it's worth repeating what Ray T. Kaczorowski said. "The bays are generally uninviting, due to extremely dense vegetation, boggy to swampy floors, and the presence of various vermin, including many species of poisonous snakes."

Uninviting to human visitors, but not to everyone. To go into a Carolina bay is to see nature at work at its complicated finest where the absence of water means life for some and death for others.

We saw anhingas spreading their wings like deities; grasshoppers perched on wildflowers; wind-twisted cypresses; monarch butterflies; delicate, deadly Venus flytraps; and *Crotalus horridus,* the canebrake rattler. We never left a bay without feeling we had been on a wilderness expedition. We came away with renewed appreciation for the intricate workings of the natural world, which some call "ecology."

When Henry David Thoreau wrote in *Walden,* "In wildness is the preservation of the world," he could have been referring to Carolina bays, those mysterious landforms that capture the imagination. Thoreau wrote, "I went to the woods because I wished to live deliberately, to front only the essential facts of life, and see if I could not learn what it had to teach, and not, when I came to die, discover that I had not lived."

Live some. Visit a Carolina bay.

Destroying Paradise

At White Lake, North Carolina, we saw a Carolina bay that was a legendary shag haunt when North and South Carolina's official state dance was ascending to glory. As for White Lake being a Carolina bay, well, it is, in name only.

I first heard of White Lake while researching the history of the shag, that dance that became a beach legend. I saw old photographs of the place and recall an oval lake with no homes, just a pavilion that jutted over clear waters. I had seen so many photographs of White Lake from yesteryear that I was unprepared for reality. Arriving at the fabled venue, I saw nothing resembling a wilderness oasis. I saw plenty of development and lots of concrete. I saw docks and piers. It was nothing less than an inland Myrtle Beach, albeit laid back. "We don't get excited about nothing here," a woman told me.

I drove around the lake. I saw a water park, a wacky putt putt golf course with an elephant standing atop a manmade waterfall, a house on stilts, a pavilion, a carnival ride, and all manner of development. I saw motel, cottage, and campground accommodations as well as permanent home sites. The Carolina bay looked like a house-fringed, water-filled limestone sink in Orlando.

And so at White lake you have this fascinating confluence: one of the larger Carolina bays in the region, a historic shag venue, and a summer destination for people who like the trappings of Myrtle Beach without having to go to the coast. When the summer vacation season arrives, the population skyrockets as people from West Virginia, Virginia, North and South Carolina, and beyond come to this Carolina bay known for its white, sandy bottom and clear waters. Swimming here is not as dangerous as the beach with its rip currents and undertow. Some, in fact, refer to this shallow lake of 1,100 acres as "The Nation's Safest Beach."

You don't have to be a beach lover or a shagger to visit White Lake. Go see first-hand what happens when a unique landform that has long baffled scientists loses its

naturalness. You won't find pitcher plants or rich biodiversity, but White Lake is not alone. White Lake joins a long list of Carolina bays that have succumbed to man's desires. As noted earlier, urbanization sacrificed Blythe Bay in Wilmington, North Carolina, for a hospital, shopping mall, and golf course.

Aiken, South Carolina, has a Carolina bay in its midst. Located near Whiskey Road, development surrounds the bay. Of the bay the Aiken city government wrote that "our bay is not undisturbed and even the bay itself was farmed for many years. Streets and other developments around it affect water flow into the bay and the City has installed pumps to keep water from overflowing into nearby neighborhoods." And then an encouraging note: "The bay supports a wide variety of wildlife and volunteers have restored the more natural Carolina Bay flora."

Nearby Sun Bay at Savannah River Site is a parking lot. The University of Georgia Ecology Laboratory began studying Rainbow Bay in 1978 as a control site to examine the ecological significance of Sun Bay, destroyed in the building of the Defense Waste Processing Facility.

Man has even used Carolina bays to treat wastewater. In South Carolina in October 1990, the Carolina Bay Natural Land Treatment System was dedicated as the Peter Horry Wildlife Preserve to serve up to thirty thousand people's wastewater treatment and disposal needs. The South Carolina Department of Natural Resources, involved in the permitting process, recommended against it. While it's widely accepted science that certain wetlands can function as water treatment systems, the bays in the Grand Strand area were not good choices for wastewater treatment. They're acidic and, as such, nutrient limited. A large influx of nutrients in the form of wastewater was likely to change completely the vegetative structure of the bay.

Highways extract a toll on Carolina bays and their wildlife. In Horry County, South Carolina, controversy attends the Carolina Bay Parkway (Highway 31) because of the wetlands and bays it runs through and near. Politics and allegations that the road's purpose was to abet development more than to relieve traffic congestion contribute to the controversy. Construction of the third phase was delayed due to environmental concerns and, in part, to a reduction in the number of wetland acres to be filled. Though named for Carolina bays, bays don't always receive media coverage in stories about the parkway.

Close by, International Drive, a dirt road that ends in Lewis Ocean Bay Heritage Preserve, also encountered opposition from environmentalists and wildlife groups who feared that expansion of this dirt road would hurt the wetlands it cuts through and their wildlife inhabitants. The plan is to extend

White Lake's Clear Water
Water like glass, but gone are pitcher plants and a rich diversity of life forms.

the dirt road and pave it. A large population of black bears frequent Lewis Ocean Bay Heritage Preserve, however, and the bears often cross International Drive. Environmentalists and wildlife authorities felt that bear tunnels should be built beneath the road that would wind alongside the preserve to protect people and bears. And that was the plan originally. Then a fire came along. County officials concluded that the population of bears had dwindled significantly following the notorious Barefoot Landing wildfire in 2009 that scorched thirty-one square miles and consumed seventy homes. Not building bear tunnels represented savings of three million dollars. Dr. Jim Luken has updated the situation as of this writing. "The base for International Drive is complete and now all that needs to be done is the paving and few more wetlands filling issues. Although I'm no engineer, I don't see how bear underpasses could have been built without overpasses (money) and, of course, if constructed, the bears would eventually find themselves underpassing into subdivisions. I think we can safely say that the destruction and fragmentation of bear habitat in Horry County continues at a rapid pace."[1]

The Allure of Fertile Lands

Scott Harder has indicated that it's well documented that a vast majority of Carolina bays have been altered or disturbed. "Main disturbances include conversion to agriculture or silviculture, where bays are often ditched and drained. Alteration of bay hydrology in these systems has degraded or destroyed habitat for many plant and animal species."[2]

Educating people to bays' unique ecological roles is paramount when it comes to staving off destruction. Johnny Stowe, DNR wildlife biologist and heritage preserve manager, and Bob Perry, DNR's Office of Environmental Programs director, worked on the mission-critical Isolated Wetlands and Carolina Bay Task Force. They champion Carolina bays' benefits. Carolina bays are important to South Carolinians and merit preservation.

Farmers long ditched and drained bays for agricultural use. Some farmers have reported that the bays' high organic content provides corn yields fifty to sixty percent higher than adjacent lands. Bays have been ditched and drained in order to harvest their timber. In the 1800s and early 1900s, swamps' difficult conditions thwarted large-scale lumbering operations, which protected cypresses. Bays, however, offered an alternative source. Lumber companies ditched bays, built plank roads over the marshy landscape, and cut and dragged logs out by horses or mules and, later, tractors. Despite decades of restoration efforts, some bays still display scars from ditching and paths where logs were pulled from the bay.

When you ditch a bay, the water flows away and the bay dries. When it dries the wetland vegetation goes away, and wildlife associated with the water and vegetation goes away as well. Plant succession takes hold, and in time trees dominate. Wholesale change arrives.

The degradation of bays has long taken place. We've built homes around them. Attached gristmills to them. Farmed them. Logged them. And then there's White Lake, which looks like a junior version of Myrtle Beach, where piers and boardwalks pierce the bay's watery membrane. The Savannah River Site has 343 Carolina bays or bay-like depression wetlands. Of these it's estimated that two-thirds were ditched or disturbed before federal occupation of the site.

In 1951, prior to the establishment of the Savannah River Plant, as it was first known, agriculture, cattle, and forestry were the chief land-use activities in and around bays. The extreme security the plant required protected the bays from disturbances, although significant portions of the periphery of Mona and Woodward Bays were clear-cut and replanted with longleaf pine in 1987. We've seen that in Barnwell County, people once grew asparagus in a bay. The list goes on.

Few Carolina bays remain in a relatively natural condition; most have been drained and cleared for agriculture, pine plantations, or housing developments. Carolina bays are considered isolated wetlands under federal law, with no inflowing or outflowing streams; thus they are not protected under the federal Clean Water Act, a nearsighted judgment. In Georgia alone twenty-five rare plant species and eighteen rare animal species are associated with Carolina bays.

Temporary Water, Permanent Benefits

A bay that has been ditched, drained, and converted to crops—soybeans, for instance—looks like another field at ground level. From high above it is a different story. If you go to Woods Bay, you'll see an exhibit that compares it to its neighboring destroyed bay, Dials Bay. Each is the same size. From the air they look a bit like alien eyes. One is green and verdant; the other bare. To see this exhibit is to see the fate of thousands of bays that fell to the saw and plow.

This aerial photograph of Woods Bay and Dials Bay reveals a tale of two bays. Woods Bay is lush and green till fall turns its cypress orange and its sand-rim oaks russet. Dials Bay, however, is crossed by manmade boundaries and row-crop furrows. It's dry and brown, though you can still see its elliptical outline. The two bays, once twins, share an oval shape only.

Spring Oaks

HORRY COUNTY, SOUTH CAROLINA

Backlight highlights the red veins of turkey oak leaves. Sand rims often host the scrubby oaks, which take well to poor, sandy soil—conditions few other oaks survive. Turkey oaks get their name from their leaves, which resemble a turkey's foot.

The diverse habitat once packed into Dials Bay is gone, and this scenario has been oft repeated in other bays. The impact ripples out and above with the loss of many wildlife species, including bats. Studies reveal that bats flying over a drained bay are far fewer than those over undisturbed bays, and certain bats are species of concern. Restore a bay's water, and the bats flying over it increase.[3]

Some organisms are 95 percent water, while most of the rest are more than 50 percent water. It doesn't take much to destroy a bay's intricate web of life. Just get rid of the water. Carolina bays provide water in areas where standing water would otherwise be a rarity. Destroy the bay, and where will some species obtain drinking water and areas to breed?

Earth's most mysterious landform, when undisturbed, teems with life, and that alone is reason enough to protect undisturbed bays, but that hasn't been the case historically. Way too many bays have succumbed to ditching and draining. And it's important to point out that preserving a bay yet clear-cutting up to its edges is damaging also. Bays provide connectivity, avenues for species to find what they need. Remove the natural buffers, and you break the connectivity.

Steve Bennett, writing in the Sierra Club of South Carolina's *Congaree Chronicle,* points out the value of "temporary" water and connectivity. "The gopher frog may be the most important species that has taught me, and many others, about habitat connectivity. Even with the greatest breeding ponds functioning just the way they are supposed to, if they are surrounded by clear-cut or fire-suppressed forest, there will be no gopher frogs. Conversely, there could be many thousands of acres of the best-managed longleaf pine forest around, but if the breeding ponds have been ditched and drained, or deepened to 'improve' fishing, there will be no gopher frogs. This also goes for many other amphibian species, especially rare ones and those declining because the environmental conditions they need to thrive are also rare and declining. This is why it is important to pay attention to our 'flagship' species and understand the true value of wetlands that aren't wet all the time—temporary water!"[4] Similarly Harder expresses his wish for Carolina bays: "Whether accomplished through mitigation banking or through conservation programs such as the Heritage Trust Program, I would like to see more Carolina Bays protected and/or restored to further enhance the natural resource heritage of our state."[5]

In South Carolina just off US Highway 78, Ditch Pond straddles the Aiken/Barnwell county line. It's not a pond; it's a twenty-five-acre Carolina bay, and it's a South Carolina Department of Natural Resources heritage preserve that will remain undisturbed. You will find ditches here—two, in fact, and the possibility exists that they adversely affect the bay's water.

Managers of preserves and protected natural areas use the acronym DFC: desired future condition. This term expresses practical and visionary considerations for an ecosystem's future. The DFC for Ditch Pond Heritage Preserve is to keep it an intact Carolina bay with self-regenerating populations of rare plant species.

What if Ditch Pond had been left severely ditched and drained, as its name implies? A ditched bay is of far less value to wildlife. A bay that stands in water much of the time is valuable, as are bays that feature water half the time. Bays that feature water at any time are paradises of their own making. Had Joni Mitchell spent time in Carolina bays, the lyrics to "Big Yellow Taxi" might have gone like this: "They drained paradise and put up a soybean crop," or "They ditched paradise and logged all the trees."

Either way you don't know what you've got until it's ruined. And a lot of Carolina bays are just that. Ruined.

Regaining Paradise

A sharp, lilting whistle echoes across the bay. *Bob, bob, white. Bob, bob, white.* That's a welcome sign of healthy habitat. Outside of Carolina bays, I never hear bobwhites like I did growing up in rural Georgia. Within five minutes of entering almost every bay I've been to, I hear the bobwhite's call.

Sportsmen and -women should like the idea of preserving bays. We all should. We've lost many bays through the years, and even in the bays we protect, people still find ways to do harm. As evidence I present the poaching of Venus flytraps. For many years poachers have canvassed bays and other areas gathering up a thousand or more plants. On the black market, they sell for a measly twenty-five cents each. That's why poachers go for huge volumes. Still that pales compared to the days when swamp phobia led people to drain bays. Drain a bay; ruin everything.

The poaching of flytraps aside, better days are here. People are reclaiming bays—regaining paradise. In North Carolina near Lumberton, the story of Warwick Mill Bay is a good one. Warwick Mill Bay is one of North Carolina's last relatively undisturbed big bays. And it should stay that way. Conservation and wildlife organizations and non-profits came together to preserve this 977-acre bay. Audubon North Carolina, North Carolina Division of Parks and Recreation, and the Conservation Fund bought the bay.

Valuable alone for its size and diversity, the bay also represents an important resource for bird conservation. It supports one of North Carolina's largest wood stork colonies. Heavy with woods, the bay provides prime nesting habitat for Central and South American migrant songbirds, such as the prothonotary warbler, northern parula, yellow-throated warbler, and others. As well native duck species frequent the bay. Water draws them in, but a bay that has been ditched and drained has no such allure. To regain paradise, we must return water to it, and that is happening.

Rainbow Beauty

ROBESON COUNTY, NORTH CAROLINA

This wood duck male showcases its rainbow-like plumage. A broad tail and short but broad wings let wood ducks fly through wooded terrain with ease. True to their name, they nest in tree cavities.

Awareness of Carolina bays' vital roles in conservation and natural history has elevated the need not only to save them but also to restore bays that people have heavily altered. And so not all is doom and gloom. South Carolina Department of Natural Resources hydrologist Scott Harder reported that "the Heritage Trust Program, currently administered by the South Carolina Department of Natural Resources, has highlighted the need to preserve and/or restore these unique ecosystems and has already placed a number of bays under protection through the program."[1]

Ditching and draining a bay impose drought on it. Vegetation dies. Aquatic organisms die. If habitat is left idle, plant succession takes over. Invasives arrive. If the bay was drained for agricultural uses, man's heavy equipment arrives, and soon it is timbered, plowed, and planted. It remains, however, a depression. The potential to hold water remains. Plug a bay, and it gets another chance at life. Water returns as rain cycles come and go. Vegetation and organisms suited to the intermittent water or standing water return. Nature's grand plan for these elliptical depressions imposes its will. Given enough time, the bay resembles what it once was.

Today wetlands mitigation helps to restore certain Carolina bays. Harder explained that "wetland mitigation involves the restoration of wetlands to offset the impacts to other wetlands resulting from development. Mitigation banks offer credits, which are sold to developers or others whose land-use activities result in unavoidable wetland impacts. Several banks in South Carolina consist of Carolina bays and restoration efforts typically consist of plugging or filling in ditches to restore natural hydrology and the removal of non-wetland species."[2] Dr. Jim Luken added, "People have plugged ditches to restore the hydrology and the bay that used to be there. The reason for doing that is to make wetland mitigation banks. If you restore the hydrology, you'd be amazed what comes back into the system."[3]

In the past man's developments have repeatedly and adversely affected waterways, pools, bogs, and other areas, harming millions of acres of wetlands and habitat. Carolina bays were not spared. Past efforts to protect natural areas proved beneficial and slowed the destruction of natural lands and their environmental functions. Protecting natural areas, however, is not enough. Restoring lost natural areas is vital.

George Howard has been in environmental mitigation since 1996. He's cofounder and chief executive officer of Restoration Systems, a company that offsets development's impact on wetlands. As his company website states, "It is less commonly understood that large-scale future improvements and net gains for the environment depend entirely on restoring previously degraded areas—rather than simply protecting the natural areas that remain."[4] Restoration Systems restores and protects land and water by purchasing a permanent conservation easement or fee-simple interest from property owners. It then restores the waterways, trees, and vegetation to duplicate the area's past natural functions.

Howard sees in mitigation and mitigation banking a new industry whose objective is restoring environmental function in one area to compensate for others' destroying preferred environment elsewhere. It offers hope for restoring Carolina bays. Just what is a wetlands mitigation bank?

The U.S. Environmental Protection Agency's sleep-inducing answer demands a cup of coffee. "A mitigation bank concerns the restoration, establishment, or enhancement of a wetland, stream, or other aquatic resource area for compensation under Section 404 of the Clean Water Act. A government agency, corporation, nonprofit organization, or other entity can create a mitigation bank under a formal agreement with a regulatory agency."

A mitigation bank comprises four components. The site is the physical acreage restored, established, enhanced, or preserved. The instrument is a formal agreement between bank owners and regulators establishing liability, performance standards, management and monitoring requirements, and approved bank credit terms. The interagency review team provides regulatory review, approval, and oversight of the bank. And the service area is the geographic area in which permitted impacts can be compensated for at a given bank.

Mitigation banking reduces doubt as to whether compensatory mitigation will offset project impacts. It provides extensive financial resources, planning, and scientific expertise not always available. It reduces permit-processing times and provides more cost-effective compensatory mitigation opportunities while better using limited agency resources in the review and monitoring of compensatory mitigation projects. Scott Harder puts it in everyday language: "Several banks in South Carolina consist of Carolina bays, and restoration efforts typically consist of plugging or filling in ditches."[5] As Jim Luken indicated, "you'd be amazed what comes back into the system."

Another proactive program is the Agricultural Conservation Easement Program, which provides financial and technical assistance to help conserve agricultural lands and wetlands and their related benefits. Under its Agricultural Land Easements component, the Nature Conservancy and U.S. Department of Agriculture's Natural Resources Conservation Service (NRCS) helps Indian tribes, state and local governments, and nongovernmental organizations protect working agricultural lands and limit nonagricultural uses of the land. Under the Wetlands Reserve Easements component, NRCS helps to restore, protect, and enhance enrolled wetlands. At one time the Wetlands Reserve Program (WRP), a voluntary program, offered landowners the opportunity to protect, restore, and enhance wetlands on their property.

As for Carolina bays, restoration is considered successful when water returns to the bay and vegetation found only in or on the water lives in it once again. Regaining paradise begins with water. "These wetlands' ability to retain water along with their varying hydroperiods provides interesting habitat conditions," observed Harder. "With their unique oval shapes and northwest to southeast orientation, Carolina bays provide valuable ecosystem functions and are worthy of our attention and protection. They add greatly to South Carolina's (all relevant states) rich natural beauty and heritage. Anybody who has been in the interior of one these bays, Cathedral Bay, for example, can attest that they offer a unique outdoor experience."[6]

Case in point: In Horry County, South Carolina, Harry McDowell enrolled 447 acres of wetlands into WRP under a permanent easement. That opened the path to one hundred percent cost-share for restoration. Restored to its original water depth, the area gave two Carolina Bays a way to return to life. Post restoration, McDowell donated the land to the Horry County Conservation Foundation, a nonprofit conservation land trust.[7] The wetland area thrives and promises to provide a lifetime of education, recreation, and enjoyment for Horry County residents.

Reviving and Protecting Bays

An uplifting story as well comes from a bay in Georgetown County, Carvers Bay. At four miles long, it's one of the larger remaining bays. During World War II bombs rained down on Carvers Bay. Drained, the bay provided a setting for tree farming as well. Better days arrived in 2012 when the late "Doc" and Hazel Young's children placed a conservation easement on 2,100 acres, about half of Carvers Bay. Within half an hour, the South Carolina Nature Conservancy bought the land from the Young Heirs Partnership. The NRCS had been working with the Young family for more than a year to protect more than half of the 4,000-acre Carvers Bay.[8]

From 1943 to 1945, the U.S. Army Air Corps flew pilots, bombardiers, and navigators over the 35,000-acre Georgetown Bombing Range. The bombing left chemicals and other problems. In 2007 the U.S. Army Corps of Engineers determined that most of the chemicals were no longer a risk. Multitudes of .50 caliber rounds had killed many trees, and of those that remained, many fell to the saw.

Today, thanks to diligence and generosity, Carvers Bay takes its place alongside more than 126,000 acres of protected lands along the rivers that feed Winyah Bay, a bay of another kind.

In South Carolina, North Carolina, and Georgia, organizations are reviving and protecting bays. The Nature Conservancy protects North Carolina's Antioch Bay, one of the better examples of clay-based Carolina bays and among the Southeast's more highly imperiled ecosystems. One of North Carolina's finer examples of the rare cypress savanna natural community, Antioch Bay supports numerous rare plants and animals, including the federally endangered Michaux's sumac and two species of special concern, Boykin's lobelia and awned meadow beauty. The site supports one of just three populations of Florida peanut-grass in North Carolina, and it supports breeding populations of at least sixteen species of amphibians. Among them you'll find rare species such as Mabee's salamander, tiger salamander, and ornate chorus frog. You'll also find other rare animals such as the glossy crayfish snake and yellow-crowned night heron, found infrequently in certain states.

In Georgia the Department of Transportation purchased Dixon Bay to protect it permanently as part of a mitigation program for wetlands affected by the building of roads.

Dixon Bay, a 334-acre property in Screven County, includes a 155-acre Carolina bay. The DOT and Georgia's Department of Natural Resources struck a memorandum of agreement that lets DNR manage the property, with DOT providing stewardship funds to assist DNR management. Dixon Bay, an ecologically valuable wetland, gives wood storks places to feed, and suitable habitat appears to exist for the Canby dropwort (*Oxypolis canbyi*).

As isolated wetlands, Carolina bays receive limited regulatory protection, which serves as a threat itself in that it doesn't discourage development. The bar must remain high when it comes to reviving and protecting bays. After all people have a vested interest in Carolina bays. One of their residents, salamanders, serves as environmental indicators. They tell us much about the health of the bays. As the health of salamanders' environment goes, so goes humankind's.

Mother Nature is resilient. Given the chance, her legions of organisms will return bays to their former glories. At the Savannah River Site, Lost Lake, a wetland in the sandhill area with no obvious source of water, came by its name in an honest way. In

Dead Wood Lover
KERSHAW COUNTY, SOUTH CAROLINA
Looking a bit like redheaded matches, British soldier lichens can be found in bays recycling dead wood. Lichens are not one single species or organism. They form when fungi and blue-green algae, or cyanobacteria, grow together in a symbiotic relationship. Together they play a role in the cycle of life.

the early 1970s, Drs. Whit Gibbons and Rebecca Sharitz of the Savannah River Ecology Laboratory were trying to find an anonymous Carolina bay close to where new SREL facilities were to be built. They forgot to take a map with them. Having had no luck finding the bay, they returned to the lab and referred to the hard-to-find bay as "Lost Lake." The name fit.

In 1950 when the Savannah River Site sealed away the public, Lost Lake had been just that—lost for some time. You could see its outline from high above, however. You could see, too, the ditch that drained away its water, preparing its conversion to farming. When the site closed, an interesting thing happened. People could no longer access the bay

and keep the ditch open. The clogged ditch lost its ability to drain water, and Lost Bay gradually returned to its natural ways. Nature reclaimed what had rightfully belonged to it.

Chances are we won't be building other Savannah River Sites. We won't cordon off square miles of land that inadvertently protect bays from disturbance. We have to protect and restore them in other ways, and, as we've seen, other ways exist.

Imagine the return of water to dry, brown ovals throughout the coastal plain. Imagine their restoration to verdant ovals where all manner of flora and fauna find shelter from the suffocating encroachment of civilization. Envision wild places where you, too, can escape civilization, places where you get the feeling you've transported yourself to Africa.

Tree Antlers

BERKELEY COUNTY, SOUTH CAROLINA

This cypress, a bit evocative of atypical antlers or the bristlecone pine, appears dead, but it isn't. It lives in hospitable conditions, unless its bay is drained. The bristlecone pine lives in regions with virtually no rainfall; cypresses, however, grow in Carolina bays, which, absent human meddling, enjoy varying levels of water and great diversities of life.

Afterword

The Future of Carolina Bays

In 1930 aerial photography introduced Carolina bays to the world. Despite the passing of so many years since then, few people know much about them. Many people don't even know they exist. It's our hope that the future brings a higher awareness of Carolina bays' worth. We would like that awareness to lead to more bays being preserved and greater numbers of bays being restored to a more natural state.

We hope that enlightenment shines a brighter light on Carolina bays, for the notion that they are annoying wet places is in no way true. They are magnificent wild places where people marvel at the workings of nature.

It's encouraging to see organizations working hard to preserve and restore Carolina bays. It's heartening to think that more bays can become preserves, wild places where future generations can glimpse what once was common. Land trusts, conservation groups, departments of natural resources, associations, nonprofits, and private citizens have done much to protect Carolina bays.

More work lies ahead. Humankind is relentless when it comes to surveying, leveling, and developing land. Eventually, as fortunes and trends change, many of these creations are abandoned. What a shame it'd be to destroy more Carolina bays just to build a plant or parking lot that time will relegate to disuse.

What created Carolina bays? Well, we hope the origin mystery endures. As smart as we all are or think we are, Earth still holds secrets. Among those secrets are how and what formed the Carolina bays. The 1930s aerials opened our eyes to the bays, and today's use of LiDAR promises more discoveries. The future may tell us just what created Carolina bays, but it's our hope that their origin remains a mystery. Were we to comprehend fully how the bays formed, they'd lose the mystique that surrounds them. Lose the mystique, and an attitude of complacency would surface, and complacency never leads to good things.

It's vital to foster an awareness of Carolina bays' uniqueness and place in the natural world. From the days in the early 1980s, when Robert and I first learned of the bays, to today, one thing has consistently amazed us: so few people know anything about them. Here's hoping our populace and leaders in government and industry understand just what a rare and beautiful thing Carolina bays are. Spread the word as to how magnificent these repositories of wildness are.

Perhaps the future holds ecotourism for select Carolina bays. How wonderful to emphasize the need for their conservation, how vital to educate people to their worth, and what a blessing to get communities to treasure their bays and participate in their preservation and restoration.

Here's hoping we see less asphalt and cement; that we see more carnivorous plants and sedges; that the future brings more wildness to our part of the world, and that the mystique of the Carolina bays fascinates future generations.

Notes

Introduction

1. Thomas Gardiner, "SRS Home to Rainbow Bay, Longest Ever Amphibian Study." *Augusta Chronicle,* April 22, 2017. (Wetlands ecologist Rebecca Sharitz first related this fact to us in a conversation on February 27, 2014.)

The Big Oval Picture

1. Janna Sasser, "Naturalists Notebook: Carolina Bays: Another Man's Treasure," *North Carolina Sea Grant Coastwatch*, Autumn 2015, https://ncseagrant.ncsu.edu/coastwatch/previous-issues/2015–2/autumn-2015/Carolina-bays-another-mans-treasure/.
2. Silvia E. Piovan and Michael E. Hodgson, "How Many Carolina Bays? An Analysis of Carolina Bays from USGS Topographic Maps at Different Scales," in "Dynamic and 3D Illustrations," special content section, *Cartography and Geographic Information Science* 44, no. 4 (2017).
3. Jerry Reynolds, UNC TV Science, Carolina Bay, "Bays of Mystery." http://science.unctv.org/content/bays-mystery.
4. Michael Tuomey, *Report on the Geology of South Carolina.* (Columbia, S.C., 1948).
5. Ibid.
6. Diane Tennant, "The Carolina Bays: Explaining a Cosmic Mystery," *Virginian-Pilot,* September 7, 2008.
7. Raymond T. Kaczorowski, "The Carolina Bays: A Comparison with Modern Lakes" (Ph.D. dissertation, Coastal Research Division Department of Geology, University of South Carolina, 1977), Technical Report No.

13-CRD. Also in "The Carolina Bays and Their Relationship to Modern Oriented Lakes" (master's thesis, University of South Carolina, 1977).
8. L. C. Glen, "Some Notes on Carolina Bays," *Science,* October 11, 1895, pp. 472.
9. C. Wyeth Cooke, *Carolina Bays and the Shapes of Eddies,* Geological Survey Professional Paper 254-I (Washington, DC: United States Government Printing Office, 1954), 195.
10. Bob Kobres, "A Nickel Pickle, Part C," *Cosmic Tusk,* July 12, 2010, https://cosmictusk.com/a-nickel-pickle-part-c/.

Amazing Landforms

1. Antonio Zamora, "A Historical Perspective of Research on Carolina Bays," *Scientific Psychic,* September 20, 2017.
2. "The First 'Bird's Eye' View of Carolina Bays," http://defendgaia.org/bobk/cbaymbsc.html (accessed April 24, 2019).
3. F. A. Melton and W. Schriever, "The Carolina 'Bays'—Are They Meteorite Scars?" *Journal of Geology* 41 (1933): 52–66.
4. Ibid.

The Great Origin Debate

1. Janna Sasser, "Naturalists Notebook: Carolina Bays: Another Man's Treasure," *North Carolina Sea Grant Coastwatch*, Autumn 2015, https://ncseagrant.ncsu.edu/coastwatch/previous-issues/2015–2/autumn-2015/Carolina-bays-another-mans-treasure/.
2. W. F. Prouty, "Carolina Bays and Their

Origin," *Geological Society of America Bulletin* 63 (1952): 167–224.

3. Raymond T. Kaczorowski, "The Carolina Bays and Their Relationship to Modern Oriented Lakes" (master's thesis, University of South Carolina, 1977).

4. Henry Savage, *The Mysterious Carolina Bays* (Columbia: University of South Carolina Press, 1982).

5. Christopher R. Moore, "Carolina Bay Formation and Evolution: Kaczorowski Was Right!," paper presented at the Geological Society of America annual meeting, Charlotte, N.C., November 4–7, 2012.

6. Andy Wood, "The Carolina Bay Enigma, One that Baffles the Understanding," *Wrightsville Beach Magazine,* March 2013.

7. Conversation with Dr. Jim Luken, Lewis Ocean Bay, Horry County, S.C., May 21, 2014.

8. "Part I: Michael Davias of Perigee Zero Presents on the Carolina Bays," YouTube video, posted by George Howard, December 18, 2009, https://www.youtube.com/watch?v=xEio7tfD2fA.

9. Paige B. Jarreau, "Alien Meteors, or Just Plain Wind and Water?," Museum Blogs, North Carolina Museum of Natural Sciences, June 4, 2013, https://ncmns.wordpress.com/2013/06/04/alien-meteors-or-just-plain-wind-and-water/.

10. George Howard, "Journeys with George: Did a Comet Cause the Carolina Bays?," blog, Restoration Systems, February 23, 2009, https://restorationsystems.com/press-rs-in-the-news/journeys-with-george-did-a-comet-cause-the-carolina-bays-2/.

11. Kaczorowski, "Carolina Bays."

12. Ibid.

Dispatches from the Field

1. Whit Gibbons, *Population Growth, Global Viewpoints* (Farmington Hills, Mich.: Greenhaven, 2009), 23.

2. "Barnwell County, South Carolina," http://www.carolana.com/SC/Counties/barnwell_county_sc.html (accessed April 24, 2019).

3. Conversation with Dr. Jim Luken, Lewis Ocean Bay, Horry County, S.C., May 21, 2014.

4. Melissa Dowland, "The Most Wonderful Plant in the World," Roads End Naturalist, March 15, 2017, https://roadsendnaturalist.

com/2017/03/15/the-most-wonderful-plant-in-the-world/.

5. Luken conversation.

6. Conversation with Linda Lee, wetlands ecologist, Thunder Bay, Savannah River Site, S.C., April 1, 2014.

7. Brett M. Moule, Manager, Management Plan for Ditch Pond Heritage Preserve, South Carolina DNR, 2007, 1, http://www.dnr.sc.gov/managed/pubs/DPHP2-26-07MP.pdf (accessed April 24, 2019).

8. Ernest Hemingway, *True at First Light* (New York: Scribner, 1999), n.p.

9. "Hypnotized," recorded by Fleetwood Mac, music and lyrics by Bob Welch, https://www.songfacts.com/facts/fleetwood-mac/hypnotized (accessed April 24, 2019).

A Square Foot of Grass

1. Scott Harder, email, January 18, 2013.

Destroying Paradise

1. Dr. Jim Luken, email, June 15, 2017.

2. Scott Harder, email, January 18, 2013.

3. Jennifer R. Kindell, *South Carolina Bat Conservation Plan* (Columbia: South Carolina Department of Natural Resources, 2018), 76.

4. Steve Bennett, "Temporary Water," *Sierra Club's Congaree Chronicle,* Spring 2017, 12.

5. Harder email.

Regaining Paradise

1. Scott Harder, email, January 18, 2013.

2. Ibid.

3. Conversation with Dr. Jim Luken, Lewis Ocean Bay, Horry County, S.C., May 21, 2014.

4. Restoration Systems, https://restorationsystems.com/.

5. Harder email.

6. Ibid.

7. Amy M. Overstreet, "Unique Wetlands Make a Comeback in Horry County," United States Department of Agriculture, Natural Resources Conservation Service, South Carolina, https://www.nrcs.usda.gov/wps/portal/nrcs/detail/sc/newsroom/stories/?cid=nrcs142p2_015637 (accessed April 24, 2019).

8. Tommy Howard, "The Nature Conservancy Protects Carvers Bay," *South Strand News,* April 17, 2012.

9. Harder email.

Online Sources of Images

"Benefits of Fire." Cal Fire, https://www.fire.ca.gov/communications/downloads/fact_sheets/TheBenefitsofFire.pdf.

"Blackberry." ailmentarium, https://www.alimentarium.org/en/knowledge/blackberry.

brainiac. "How To Find Frog Eggs." Sciencing, April 24, 2017, https://sciencing.com/find-frog-eggs-2184532.html.

"Garden Spiders: "Facts, Identification and Control." Orkin, https://www.orkin.com/other/spiders/garden-spiders/.

"How Alligators Work." How Stuff Works, https://animals.howstuffworks.com/reptiles/alligator2.htm.

"How Much Water Is in the Atmosphere?" The Why Files: The Science behind the News, April 19, 2010, https://whyfiles.org/2010/how-much-water-is-in-the-atmosphere/index.html.

"Lake Waccamaw State Park." North Carolina State Parks, https://www.ncparks.gov/lake-waccamaw-state-park.

"Lavender Leaves, Golden Pollen." Wildflowers of the United States, https://uswildflowers.com/detail.php?SName=Rhexia%20mariana.

"The Most Wonderful Plant in the World." Harvard Forest, https://harvardforest.fas.harvard.edu/sites/harvardforest.fas.harvard.edu/files/frankMortonJones/pdf%20files/Wonderful%20Plant.pdf.

"Mourning Dove Identification." All about Birds, Cornell Lab of Ornithology, https://www.allaboutbirds.org/guide/Mourning_Dove/id.

"Northern Cardinal Sounds." All about Birds, Cornell Lab of Ornithology, https://www.allaboutbirds.org/guide/Northern_Cardinal/sounds#.

Parry, Wynne. "How the Venus Flytrap kills and Digests Its Prey." Live Science, September 5, 2011, https://www.livescience.com/15910-venus-flytrap-carnivorous.html.

"*Sarracenia rubra*." Botanical Society of America, https://botany.org/Carnivorous_Plants/Sarracenia_rubra.php.

"Spotted Salamander (*Ambystoma maculatum*)." Savannah River Ecology Lab, University of Georgia, https://srelherp.uga.edu/salamanders/ambmac.htm.

"Sundews." National Wildlife Federation, https://www.nwf.org/Educational-Resources/Wildlife-Guide/Plants-and-Fungi/Sundews.

"Toothache Grass." US Department of Agriculture, Natural Resources Conservation Service Plant Fact Sheet, U.S. Department of Agriculture, https://plants.usda.gov/factsheet/pdf/fs_ctar.pdf.

Volk, Tom, and Sean Westmoreland. "Tom Volk's Fungus of the Month for January, 2003," https://botit.botany.wisc.edu/toms_fungi/jan2003.html.

"Welcome to the Savannah River Site Deer Control Activities." Savannah River Site, https://www.srs.gov/general/deer_hunt/hunt.htm.

"What Are Lichens?" U.S. Department of Agriculture Forest Service, https://www.fs.fed.us/wildflowers/beauty/lichens/whatare.shtml.

"What Is a Pocosin?" National Ocean Service, https://oceanservice.noaa.gov/facts/pocosin.html.

"What Is LiDAR?" National Ocean Service, https://oceanservice.noaa.gov/facts/lidar.html.

"What Is That Big Orange Spider?" The Infinite Spider, A Science and Nature Blog for Naturalists and Educators, https://infinitespider.com/big-orange-spider/.

"Yellow Fringed Orchid." National Resources Conservation Service Plant Guide, U.S. Department of Agriculture, https://plants.usda.gov/plantguide/pdf/cs_plci2.pdf.

Index